THE MYTHS ABOUT NUTRITION SCIENCE

Many nutrition science and food production myths and misconceptions dominate the health and fitness field, and many athletes and active consumers unknowingly embrace a myriad of what can be deemed "junk science" which has now infiltrated many related science fields. Consumers simply have no reliable source to help them navigate through all the hype and fabrication, leaving them vulnerable to exploitation.

The aim of *The Myths About Nutrition Science* is, then, to address the quagmire of misinformation which is so pervasive in this area. This will enable the reader to make more objective, science-based lifestyle choices, as well as physical training or developmental decisions. The book also enables the reader to develop the necessary critical thinking skills to better evaluate the reliability of the purported "science" as reported in the media and health-related magazines or publications.

The Myths About Nutrition Science provides an authoritative yet readily understandable overview of the common misunderstandings that are commonplace within consumer and athlete communities regarding the food production process and nutrition science, which may affect their physical development, performance, and long-term health.

David Lightsey is a Food and Nutrition Science Advisor to Quackwatch.org. He has been combating health misinformation in the health marketplace for 31 years, with the non-profit consumer advocacy organization, National Council Against Health Fraud until 2011, as well as with Quackwatch.org (www.quackwatch. org). He has appeared on *NBC Dateline* (Hype In A Bottle—1996), regarding the deceptive marketing methods used by the sport supplement industry to exploit and fleece athletes, as well as *CBS Evening News* (2003), regarding the designer steroid issue. The peer review of his first book, *Muscles Speed and Lies—What the*

Sport Supplement Industry Does Not Want Athletes or Consumers to Know, from the American College of Sports Medicine, stated "this book is essential reading for the interested athlete or coach. Particularly, it should be required reading for those studying in the field of clinical sports science (undergraduates, graduate students), allied health professionals (ATCs, personal trainers, and dietitians), and primary care sports medicine physicians." The International Olympic Committee chose this book to be included in "the most important bibliography of the holdings of the IOC Medical Commission Collection of Sports Medicine for reference texts and scientific journals published until 2010." Other examples of where his consumer education work has appeared include peer review journals such as the *British Journal of Cancer, Lancet, American Journal of Emergency Medicine, National Strength and Conditioning Association Journal*, and *PLOS One*, as well as many national media outlets.

THE MYTHS ABOUT NUTRITION SCIENCE

David Lightsey

Routledge
Taylor & Francis Group

NEW YORK AND LONDON

First published 2020
by Routledge
52 Vanderbilt Avenue, New York, NY 10017

and by Routledge
2 Park Square, Milton Park, Abingdon, Oxon, OX14 4RN

Routledge is an imprint of the Taylor & Francis Group, an informa business

© 2020 Taylor & Francis

The right of David Lightsey to be identified as author of this work
has been asserted by him in accordance with sections 77 and 78 of the
Copyright, Designs and Patents Act 1988.

Trademark notice: Product or corporate names may be trademarks or
registered trademarks, and are used only for identification and explanation
without intent to infringe.

Library of Congress Cataloging-in-Publication Data
A catalog record for this book has been requested

ISBN: 978-0-367-31334-0 (hbk)
ISBN: 978-0-367-35429-9 (pbk)
ISBN: 978-0-429-33133-6 (ebk)

Typeset in Bembo
by Apex CoVantage, LLC

CONTENTS

FIGURES

TABLES

ACKNOWLEDGEMENTS

In addition to the obvious gratitude the author has to David Varley at Routledge, who recognized the need for this book, I would also like to thank the editing assistance of Alexandria Tebbets. Finally, I would like to thank the many other professionals who are mentioned in the book who took the time over the years to respond to my inquiries.

1

THE CONSUMER/ATHLETE'S SOURCE OF NUTRITION INFORMATION

Who Is Reliable?

Counterfeit science has become prevalent in all areas of the sciences, and it can be argued that unreliable information plagues the nutritional sciences more than most. Nutrition "science" has become so contradictory that one must learn to take every new "study" which declares to enlighten us about some purported nutritional health threat or benefit with a large grain of salt.

There are several factors why most athletes and active consumers find the process of acquiring good information in this area so problematic, contradictory, and confusing. First, the average athlete or active consumer gives too much credibility to the advice provided by unqualified sources, such as the media, which will be discussed separately in Chapter 2, as well as friends, the supplement industry, gym personnel, magazine articles, much of the internet, books, and personal trainers, as well as most coaches. The second factor are those who seek advice from those who should be good resources—such as certified athletic trainers, certified strength and conditioning coaches, and exercise physiologists, as well as registered dietitians—who may in fact not be good sources of information due to the lack of professional training from their respective programs of study, as well as failing to pursuing continuing education from appropriate sources.

Following, I will briefly discuss the problems and value of each potential resource of nutrition science information.

Friends

This resource is essentially the blind leading the blind. Unless the friend happens to have a formal degree in the area, the athlete seeking information from a friend is highly unlikely to receive it, but instead, obtain potentially dangerous information, especially among teenage males. As an example, several years ago, I had a student attending my class, who, during his senior year in high school, was being recruited by several Division I college football programs as a running back. This all changed one afternoon during his senior year while working out with his teammates in the weight room. He had already ingested what he had stated was his pre-workout drink which had already "amped him up quite a bit." Regardless, he succumbed to the advice of his friends he was training with and tried several other versions of similar products, which soon after led to a heart attack in the gym before completing his workout, and then a stroke on the way to the hospital. This obviously put an end to his football pursuit. The details of his blood pressure at the time of the incident were not available, but there is certainly a logical assumption which can be made: that the resulting high blood pressure from the supplements, in combination with the heavy lifting, played a significant role in the incident.

Here is a bit of advice for those reading this book: many teenage males are making lifestyle decisions in environments where you have a combination of raging testosterone, undeveloped intelligence, no wisdom, and the desire to assume any risk, believing they are immune from the potential consequences. Combine that with often no adult male supervision, and this can be summed up as a disaster just waiting to happen. As a result, teenage male athletes are far too likely to pursue the boundaries of behavior which can be potentially disastrous to themselves or others, so they are not a wise choice for seeking guidance from.

The Supplement Industry

The supplement industry, for the most part, is a modern version of a snake oil salesman. *The Cambridge Business English Dictionary* describes a snake oil salesman as "someone who deceives people in order to get money from them."[1] Wikiquote describes a snake oil salesman as "someone who knowingly sells fraudulent goods or who is himself or herself a fraud, quack, charlatan, and the like." *The Free Dictionary* describes a snake oil salesman as "someone who sells, promotes, or is a general proponent of some valueless or fraudulent cure, remedy, or solution." Just take your pick, because they all apply to most of the supplement industry. If the supplement industry relied on the facts and truth, most of its market would not exist.

From 1988–1990, Jerry Attaway, M.Ed., who at the time was the physical development coordinator for the San Francisco 49ers, and I worked on a project for the National Council Against Health Fraud, addressing the deceptive marketing methods used by the sports supplement industry to fleece and exploit the typical misunderstandings of most athletes.

The investigation resulted in the publication of a paper with the same title in the National Strength and Conditioning Association Journal in April 1992 entitled "Deceptive Tactics Used in Marketing Purported Ergogenic Aids." This paper, as well as other work I was doing at the time, led to a 1996 appearance on *Dateline NBC* called "Hype in a Bottle," as well as a *CBS Evening News* appearance regarding the steroid issue in Major League Baseball, and later a book called *Muscles, Speed, and Lies—What the Sport Supplement Industry Does Not Want Athletes to Know* in 2006. All this highlighted the fact that for the supplement industry to survive, it must develop a fabricated market for most of its products, and to do so requires deceptive marketing methods, which misinforms consumers rather than informs them. These practices include, but are not limited to, the following:

- Misrepresenting clinical studies or taking them out of context.
- False, exaggerated, or purchased endorsements.
- Unreliable testimonials often provided by athletes who are clearly using steroids.
- Patent misrepresentation. Patents do not indicate the product has been proven to be effective or safe, just different, which most consumers misunderstand.
- False advertising.
- Fabricating research or omitting relevant facts.
- Stating the product has been university tested when in fact it has not been.
- Claiming their research is not for public review due to the "proprietary" blend of ingredients.

Before considering advice from the supplement industry, consider the wisdom that there are certain situations where "no advice is better than bad advice." I believe this phrase applies to this industry. The liability of supplements chapter will further illustrate this point.

Gyms and Personal Trainers

Gyms are essentially an extension of the supplement industry. They often rely on in-house supplement sales to enhance their profit margins by utilizing the same misinformation developed by the supplement industry. Gym staff are largely poorly educated in this area and have a limited science background. Personal trainers who are only "certified" lack the understanding necessary to be a reliable source. In 2013, researchers from Florida International University published "Sports Nutrition Knowledge and Practices of Personal Trainers" in the journal *Community Medicine & Health Education*.[2] They assessed the sports nutrition knowledge and practices of 60 personal trainers from 14 different fitness centers in South Florida, as well as 69 personal trainers at the annual National Strength and Conditioning Associations conference. They found that the average knowledge score of personal trainers was 59.7%. Now, if you were a student of mine, I would of course round this percentage up to your advantage for a "D" grade in the course, but either way, this individual would not be someone I would seek out for guidance in this area. Now, I am not embracing this research study from Florida International University as my only source of reasoning for my position on personal trainers. It only supports what I have heard personally from this population group for 30 years, many of whom have been my own students.

Magazines

As with the general media, magazines face the same associated problems of unqualified journalists with no science background attempting to write informative articles on subjects they may have little to no understanding of. This is not to say that there are not journalists who can produce good work in this area, which there are, but, as with the media in general, this is far more of an infrequent occurrence than a common occurrence: uncommon enough to make most magazines an unreliable source. Magazines can produce good work, but the reader essentially needs a degree in the field to know when the information is reliable, as I state in Chapter 2. However, even when the journalist has a background in science, it does not always guarantee good investigative journalism or the prevention of fabricated, fearful headlines which exaggerate what the raw data states.

Let me illustrate this issue using what is considered by most consumers as a very reliable source of objective information on many subjects: *Consumer Reports* (*CR*). Unlike my point made previously, whereas most magazines utilize journalists with no science background, the three following examples of shoddy science journalism comes from writers with science backgrounds. The first article was produced by *CR* staff and the remaining two by specific *CR* staff who are reported to be "scientist turned journalist."

On September 17, 2017, *CR* published the article "5 Vegetables That Are Healthier Cooked."[3] *CR* tries to make the case that five vegetables—carrots, mushrooms, spinach, asparagus, and tomatoes (which botanically is a fruit)—are better for you if cooked vs. eaten raw. *CR* initially states that the tips they provide will "unleash their full potential in terms of nutrition." Is this statement true, or is it just another spin on an irrelevant issue to attract readers?

> Carrots: CR states "cooking ignites this veggie's cancer-fighting carotenoids," by increasing the "concentration of carotenoids by 14 percent." First, it is true that carotenoids are one of the many plant chemicals which are associated with reduced cancer rates among those who consume them. However, this is an association, not a cause-and-effect relationship. It is the synergistic effect of the many thousands of plant chemicals that appear to be responsible for this effect, and not the isolation of any specific one, as I will repeat many times in this book. To state that just because cooking increases the concentrations of carotenoids from carrots will "ignite" carrots' cancer-fighting potential is a spin. Obtaining more of any plant chemical does not necessarily equate with improved health. The carrot already provides more than enough carotenoids in any state of ingestion. Just because you

purportedly ingest 14% more is meaningless. More does not mean better; it's just more.

Mushrooms: *CR* states "a cup of cooked white mushrooms has about twice as much muscle-building potassium, heart-healthy niacin, immune-boosting zinc, and bone-strengthening magnesium as a cup of raw ones." This statement can almost qualify as a bad riddle. Before you continue reading, stop for a moment and re-read what *CR* just stated. What is the glaring problem with this statement? It is related to one simple word: cup. *CR* really is comparing the nutrient content of a cup of cooked mushrooms, which has likely four times the number of mushrooms per cup, to a cup of raw mushrooms? What *CR* should have done was compare the actual number of mushrooms cooked vs. raw, not the volume of them. A mushroom is 92% water by weight, so when you cook them, the volume is significantly reduced—so of course a cup of cooked mushrooms will have substantially more nutrients. This is common sense.

Spinach: *CR* states, "The leafy green is packed with nutrients [true], but you'll absorb more calcium and iron if you eat it cooked." This is blamed on the oxalic acid in spinach. Oxalic acid is commonly believed to bind with both minerals and prevent their absorption. However, this would only apply to the two minerals contained in the spinach and not from other food sources you may be having with your meal, such as milk for the calcium and any meat item or beans for your iron. Additionally, there are some data indicating that the oxalic acid in spinach may not actually prevent the iron absorption. A study done in 2008, using iron isotope absorption in humans, concluded, "Our results strongly suggest that oxalic acid in plant foods does not inhibit iron absorption." This study was published in the *European Journal of Clinical Nutrition* and conducted at Institute of Food Science and Nutrition, ETH Zurich, Zurich, Switzerland.[4]

Asparagus: *CR* states, "Cooking these stalks raised the level of six nutrients, including cancer-fighting antioxidants." The value of antioxidants individually has been overemphasized for well over a decade. They are important, but all plants contain them, and they are readily supplied by any plant-based diet, cooked or otherwise. Antioxidants are not the magical, super-chemical compound consumers have been led to believe, even though they have become a very popular buzzword for marketing products (see Chapter 8). It is the mix of compounds, not a specific compound, which are related to the benefits of a plant-based diet. You also increase the production of your own antioxidants when you exercise.

CR also states cooking asparagus "more than doubled the level of two types of phenolic acid, which some studies *linked* [my emphasis] to lower cancer rates." Anytime you see the phrases "linked," or "associated," in any media article claiming a compound is either bad or good for you, significant skepticism is warranted. A "link," or an "association," has nothing to do with cause and effect, which I will state many times throughout the book. I can cherry-pick any one of the thousands of chemical compounds found in produce or grains and state the same thing. *CR* should understand the obvious issue here. Diets high in phenolic acid simply illustrates those individuals are consuming a plant-based diet, which of course will reduce their cancer rates, and not attempt to embellish the obvious or turn cooked asparagus into a magical food. This is something Dr. Oz might do, but should not be part of a *CR* protocol for reporting nutrition information. Again, just because you doubled the concentration of one of the thousands of phytochemicals found in produce by cooking it, this has nothing to do with making the food item better for you. Again, more is not necessarily better, just more.

Tomatoes: *CR* states "heat increases a phytochemical, lycopene, that has been *linked* [my emphasis] to lower cancer rates and heart disease." This is taken out of context. The "link" of lycopene to reduced rates of cancer or heart disease is not directly related to lycopene itself, but to what increased lycopene intake is representative of. A high lycopene intake is reflective of a plant-based diet, which contains literally thousands of phytochemicals playing a role in our health. The overall plant-based diet, as stated above, or the mix of the phytochemicals in them, is what is reducing the incidence of cancer and heart disease, not the lycopene itself. If this concept were true, then lycopene supplements would be in order, which has not been demonstrated in well-designed studies. So, to imply that just because you consume greater quantities of lycopene by cooking your tomatoes vs. eating them raw will provide any significant health advantage is unwarranted.

This standard bearer of consumer education receives an F grade on this report. Cooking does not "ignite" these vegetables into superfoods. They are excellent for you, regardless of how you consume them.

Other examples of *Consumer Reports* missteps include:

- A July 17, 2017 scary article headline "Low-Calorie Sweeteners May Contribute to Weight Gain—A New Study Suggests That Health Risks of Sugar Substitutes May Outweigh Their Benefits."[5]
- A May 24, 2017 article, "The Mounting Evidence Against Diet Sodas—Studies Suggest Possible **Links** [my emphasis] Between Low-Calorie Beverages and Health Risks, Though More Research Is Needed."[6]

Both are association studies which "link" or "associate" one of many possibly variables creating the cause-and-effect relationship discussed in the articles. Both articles contain all the standard pitfalls in determining a cause-and-effect relationship as I discuss next in Chapter 2. For more details of *CR* missteps regarding these articles, see Chapter 5, as well as the review of aspartame and related sweeteners in Chapter 3. The bottom line here is that *CR* has the means to produce informative and reliable information for consumers, but often fails to do so.

Books

Books can be an excellent avenue for pursuing more information in any area. However, celebrity status may guarantee sales and profits for the publisher, but rarely indicates that the information provided is credible enough to use. This is epitomized with books that are written by those with a celebrity status, since this almost guarantees the publisher will profit. Below, I will provide two examples of this using Tom Brady's book *The TB12 Method* and Dr. Oz's book *Food Can Fix It*.

I will also provide a third example using the college level introductory health textbook *An Invitation to Health, Your Life, Your Future*, published by the academic publisher Cengage, which happens to be the same publisher which produces the college beginning nutrition text I use in my course.

Tom Brady's Book The TB12 Method

Yes, Tom Brady is a master at what he does, and any young aspiring quarterback would benefit from his coaching of the position. However, his nutritional advice is something to avoid, based upon the information he has provided in his book. For the past 30 years, I have become very familiar with three things regarding sports nutrition issues. One, how misinformed most athletes are regarding nutrition science issues, even at the professional level. Two, how gullible and easy they are to exploit. Three, that there is no shortage of those willing to exploit athletes' and consumers' misunderstandings, and fleece them. Tom Brady's book is a perfect example of all points.

Outside of a little bit of common sense, most of what is contained in the nutrition chapter of the book is utter nonsense and illustrates an extreme level of nutrition illiteracy. Mr. Brady clarifies that the nutritional regimen he follows is a mix of Eastern and Western philosophies. Indeed, it is, because it certainly has nothing to do with science. Brady embraces a significant amount of misinformation regarding some very basic biology, nutrition science, the food supply process, common agricultural practices, and *the Principle of Toxicology*. His nutrition advice off the field should be ignored. It is unfortunate that due to his celebrity status, his book and the nutrition misinformation it contains will negatively influence a significant number of gullible consumers. At one time it was even a *New York Times* bestseller.

There are many errors in Brady's book, making it impossible for me to cover them all, but I will provide a sampling to illustrate that the nutritional advice in this book cannot be recommended. Here is a sampling of the misinformation:

His preference for alkaline foods: The body maintains a very limited range of pH and has various mechanisms at its disposal to do so. There is literally nothing you can consume by mouth that is going to change this. Alkaline water or foods and their relation to your body's pH is zero. This is basic chemistry.

His fear of additives and preservatives (chemophobia): This is a classic fear, of course, and only reflects a total ignorance of how we all benefit from these compounds, which I explain elsewhere in this book (see Chapter 3). Mr. Brady's misunderstanding of these issues illustrates one of the ongoing problems with consumer ignorance of the food supply process. We have 1–2% of the population growing our food, leaving roughly 98% of the population totally disconnected from the process, yet willing to complain about issues they have no understanding of.

His preference for organics: Mr. Brady embraces all the standard misconceptions about the completely fabricated organic food market, as I illustrate elsewhere in this book (see Chapter 4). Organics are neither safer nor healthier for you; nor are they better for the environment. It is biologically impossible for organics to be nutritionally superior if grown under similar conditions.

A plant's vitamin and phytochemical content is related to its genetics, and the mineral content is related to the composition of the soil and the appropriate pH it was grown in. Growers test their fields yearly for optimal mineral concentrations to make sure the soil conditions are optimal for growing. This is something every home gardener can do, as well, using a locally equipped nursery or gardener's supply store. Growers also utilize what is called a petiole test of the plant tissue during growing periods to assess that the plant is obtaining what it needs from the soil. Consumers would reject produce grown in mineral-deficient soil based upon its poor physical appearance. As an example, consider the three examples below. Image 1.1 is a tomato grown in calcium-deficient soil, Image 1.2 is citrus grown in potassium-deficient soil, and Image 1.3 is lettuce grown in phosphorus-deficient soil.

IMAGE 1.1 Tomato grown in calcium-deficient soil.

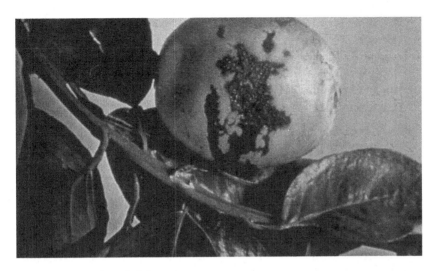

IMAGE 1.2 Potassium-deficient citrus.

IMAGE 1.3 Phosphorus-deficient lettuce.

He prefers raw vs. cooked veggies: Mr. Brady feels most veggies should be eaten raw due to the nutrient losses when cooked. Apparently, he did not read the *Consumer Reports* take on this issue. This is very misleading at best. It is true that some nutrients are unstable or lost when the food item is cooked, such as vitamin C. However, as the January 16, 2017 *Berkeley Wellness Letter* points out, "some nutrients and potentially beneficial plant compounds are less available to the body in the raw state. Heat is needed to break down plant's cell walls and release the compounds."[7] As an example, the loss of possibly half of the vitamin C due to cooking becomes a moot point when you consider the following three points:

1. How much vitamin C is stored in the body? Answer: roughly 1,500mg.
2. The three homeostatic mechanisms which help to maintain nutrient balance over a broad intake, which is discussed in Chapter 7.
3. The extensive availability of vitamin C in other foods, such as fruit, you would never consider cooking, which would be more than enough to make up the difference lost from cooked vegetables.

Finally, many consumers find cooked vegetables far more palatable when cooked vs. when eaten raw, and are more likely to eat them cooked rather than avoiding them when raw.

His bias toward locally grown food: I live in Kern County, California, the most productive agricultural area in the country, so I certainly have a wide variety of foods grown locally I could survive on. However, this is not the case for most people. Most rely on the advances in preservatives and additives to extend the shelf life and shipping distances of many products, allowing those who live far removed from "locally" grown produce to eat them in the first place.

In his book, he states, "eat as much real, organic, and local food as you can, their nutrient content is much higher than the foods you find in the supermarket."[8] This is grossly exaggerated. Advances in food technology, packaging, storage, shipping, etc., have allowed us to transport high-quality foods to almost any part of the world with more than enough nutrient content to sustain all physiological needs. Chapter 7, which illustrates the three biological mechanisms which maintain nutrient homeostasis over a broad range of intake, will help the reader further understand this issue.

As an example, I love bananas, something Brady frowns upon, but the closest banana plantation to me is 2,697 miles away in Guatemala. Also, there's a little bit of hypocrisy here. As part of his pitch for the products he is trying to sell, he states, "As part of the TB12 Method, I've also created a line of healthy snacks and protein bars."[9] How "local" are these products? Why not just tell me to eat some fruit as a snack and consume more eggs or milk?

His belief in warm-weather vs. cold-weather vs. neutral foods: This is one of the many bizarre things Brady states in his book. Apparently, his nutrition philosophy includes claiming that the outside temperature or the time of year should dictate what we eat.

Try to figure out this statement: "A few warm-weather foods that cool the body include cucumbers, asparagus, avocados, broccoli, and celery."[10] I have read a great deal of physical training recovery studies and discuss the topic during my lectures at coaches' clinics. But I have never heard of or suggested having coaches tell their athletes to eat a cucumber to assist cooling. Additionally, broccoli contains the indigestible sugar rabbinose, which the bacteria in your intestines ferment, producing gas—something no quarterback would appreciate his center having during a game. So, I am a little confused here. How can a food that induces the production of intestinal gas cool me off?

He avoids dairy: Brady feels dairy protein increases inflammation (it doesn't) "in both the digestive tract and the thyroid gland, which means your body is less able to absorb the right nutrients."[11] I find it rather odd that the very dairy proteins he is referring to (whey) are the same proteins contained in his protein supplement he would like to sell you. On the back label of his product, it states, "whey protein isolate (milk)."[12] It is also peculiar that he wants you to avoid dairy but would like you to purchase his vitamin D supplement. Just drink the milk or stand out in the sun for a few minutes.

He avoids salt: Too much salt "elevates blood pressure and interferes with our ability to eliminate toxins and waste from our cells."[13] If you're genetically salt sensitive, excessive salt can be an issue for blood pressure. However, most people are not, and salting various foods to make them more palatable is fine. For most of us, excessive salt is simply excreted through the kidneys. There is not a chance I am not going to salt my baked potato, which is often a white one, which Brady also says to avoid. Regarding the elimination of toxins issue: there is no physiological justification for this comment. The body has a multitude of ways to eliminate toxins, as I discuss in Chapter 10, and salting your food will not interfere with any of them.

He limits nightshades: Brady advises limiting mushrooms, eggplant, potatoes, strawberries, and bell peppers. Again, more nonsense. These foods are loaded with nutrients, and there is no reason to avoid them.

Combining certain foods: "Avoid eating proteins like meat, poultry, fish, or dairy with carbohydrates like potatoes, breads, wheat, or grain products."[14] This is purportedly due to the differences in pH environment needed to digest them. Again, there is no physiological basis for this belief. The stomach's pH accommodates the digestion of protein and how as soon as food enters the small intestines, the pancreas secretes the necessary bicarbonate and enzymes to digest the rest. Again, this is all basic nutrition taught to freshman college students with no science background. Can you imagine a steak with no baked potato?

He thinks you should only drink a small amount of water with a meal: Mr. Brady states that "Drinking water with meals can interfere with good digestion. I recommend only a little bit of water in order to ensure proper digestion."[15] This does not occur. The digestive tract secretes a variety of enzymes for the breakdown of protein, fats, and carbohydrates, along with water to aid the digestive process by helping to break down food particles. Water will also help prevent constipation by adding more moisture to stools.

He of course sells supplements: He states that "Even if you eat fresh, organically grown food at every meal, in can be tricky to meet your nutritional needs."[16] So, regardless of how well you follow his nutritional advice, you are still going to have to buy his supplements.

Antioxidant supplements: Brady states, "They protect the body from damage caused by free radicals."[17] However, he forgets to inform you that free radicals play a major beneficial role under most circumstances during normal metabolism. As an athlete, he should have also understood that those free radicals he is trying to suppress play a significant role as a signaling mechanism after exercise to trigger adaptation to the stress load given (see Chapter 8 for more details).

Conclusion: Most of Tom Brady's nutritional advice in the book is bizarre and without scientific merit.

Dr. Oz Book Food Can Fix It

This 2017 bestseller book is Dr. Oz's take on various "superfoods" and dietary habits, which he believes will help prevent a myriad of health conditions.

On the inside jacket of the book, it states, "Mehmet Oz, M.D., America's #1 authority on health and well-being, explains how to harness the healing power of food." Apparently, the publisher of the book was unaware of the 2014 Senate Commerce Subcommittee Hearing on Consumer Protection on Weight-Loss Advertising. Dr. Oz was also a guest of this hearing, but likely wished he had not been. During this hearing, Democratic Senator Claire McCaskill from Missouri stated, "The scientific community is almost **monolithic against you** [my emphasis] in terms of the efficacy of a few products that you have called miracles." Senator McCaskill added, "I just don't understand why you need to go there. . . . You are being made an example of today because of the power you have in this space."

Science-Based Medicine (SBM), *Dr. Oz and The Terrible, Horrible, No Good, Very Bad Day*,[18] provides this: "Oz can give good advice, but he regularly combines it with questionable statements and pseudoscience in a way that the casual viewer can't distinguish between the science and the fiction."[19] SBM also states, "Oz routinely and consistently gives questionable health advice, particularly when it comes to weight loss products,"[20] and, Dr. Oz "chooses to ignore science in favor of hyperbole. It's the antithesis of what a health professional should be doing."[21] So, the question becomes, has Dr. Oz learned from his past experiences and negative reviews of his interpretation of science to provide a book worth recommending?

Unfortunately, his book tends to follow the same pattern as his TV show, and he has apparently not heeded the advice of his peers. The book contains some good commonsense suggestions mixed with unsubstantiated nonsense. Here are some examples of his misinformation:

Recommends a daily supplement: "to ensure that you get the daily recommended amounts of vitamins and minerals."[22] He appears to have no understanding of three distinct mechanisms, which maintain nutrient homeostasis over a broad range of intakes as discussed in Chapter 7. These three mechanisms help to maintain normal physiological functions for several weeks when appropriate food choices are not available. It is Dr. Oz's possible misunderstanding of the RDAs that you must meet them daily to stay healthy. The RDAs maintain maximum reserve of a nutrient, and you do not have to have maximum reserve for normal function. Again, see Chapter 7 for more details.

The typical red and processed meat scare: Dr. Oz states, "The most recent data suggest that red meat (particularly processed red meats, like sausage and bacon), are *associated* [my emphasis] with increased heart disease, stroke, and cancer deaths."[23] He fails to understand the difference between an association vs. a cause-and-effect relationship, as I discuss in the next chapter as well as previously. This is something someone with as much medical training as he has had should fully grasp.

He bases part of this fear of red meat on the presence of L-carnitine in red meat, which is a naturally occurring compound your body synthesizes for proper metabolism of fatty acids into energy, ironically, for the heart. Dr. Oz states L-carnitine "messes with your gut bacteria."[24] This is nonsense and is related to a study published in *Nature Medicine* and performed by The Cleveland Clinic in 2013 using mice. Mice are not little people. Their physiology and how they metabolize any compound can be dramatically different than humans. In 2013, the *Mayo Clinic Proceedings* conducted a meta-analysis of 13 well-controlled studies representing 3,600 humans, not mice, and arrived at the opposite conclusion. To attempt to persuade the consumer to diminish the intake of red meat due to its content of L-carnitine is fear-mongering.[25]

The standard demonization of artificial sweeteners: "Another reason to skip diet sodas: A recent study found that drinking them is *associated* [my emphasis] with a higher risk of both stroke and dementia."[26] Again, same problem as before, he fails to grasp some basic science (see both Chapter 2 as well as Chapter 3 for additional details regarding artificial sweeteners).

Nitrates: He unnecessarily fears nitrates. Nitrates are widespread in vegetables and other foods naturally, as well as readily produced by your own saliva. See Chapter 3 for more details as to why this is a fallacy.

He loves a good colon cleanse: "The benefit to all-liquid diet is that it reduces the load on your intestines giving them a short respite."[27] My intestines get a breather between meals, nothing more and nothing less. His concept here has no basis in science. Your intestines are designed to handle food traffic as needed. Fasting or liquid meals have zero relationship to keeping your intestines healthy. He also states, "that's what dietary cleanses do, they help to remove the grit and grime from your innards, reboot all your systems, and set you up to speed through the green lights of life."[28] Embellishing the facts does not make him believable. This statement is completely irrational. There is no evidence your bowels need "cleansing"; your gut bacteria and fiber in your diet do this for you, as well as the normal movement of food waste. Additionally, I am sure a good dose of caffeine works considerably better "rebooting my systems" than some colon cleanse. It certainly sounds more enjoyable and far less potentially problematic while "speed[ing] through the green lights of life," whatever that means. For an extended review of this topic, see "Detox: What 'They' Don't Want You to Know" at Science-Based Medicine.[29]

Conclusion: Dr. Oz's 2017 book provides some good information, but you literally must have a degree in the field to separate his nonsense from the commonsense, just as I stated regarding magazine sources of science news. This is true for most books which are sold in this advice category. To support my position regarding Dr. Oz's track record for inaccurate health advice, in 2014, *The British Medical Journal* published "Televised Medical Talk Shows—What They Recommend and the Evidence to Support Their Recommendations."[30] The objective of the study was "to determine the quality of health recommendations and claims made on popular medical talk shows." The researchers randomly selected 40 episodes of each of *The Dr. Oz Show* and *The Doctors* from early 2013 and identified and evaluated all the recommendations made on each program. In the conclusions section of the study, the authors state, "approximately half of the recommendations have either no evidence [see Chapter 2 for an example of this] or are contradicted by the best available evidence. Potential conflicts of interest are rarely addressed. The public should be skeptical about recommendations made on medical talk shows."

College-Level Health Textbook

In my college course, I utilize a nutrition text published by a prominent well-respected publisher, Cengage, which, with some exceptions, provides accurate information. However, the same college textbook publisher also distributes a college-level beginning health textbook, *An Invitation to Health, Your Life, Your Future*, which errs in several areas. These errors cannot be attributed to just differences in professional opinion. Here are some examples from the newly released 18th edition:[31]

- On page 102, when referring to the B vitamins and vitamin C, "they must be replaced daily," due to their lack of storage. This is a common fallacy, as I illustrate extensively in Chapter 7. As an example, vitamin C has a reserve capacity of 1,500mg which would diminish by only 3% per day with no intake.
- On page 113, the author states "you can avoid exposure to pesticides and other chemicals by opting for certified organic foods." This is a complete misunderstanding of this issue. See Chapter 4 for an extensive review of this topic. Organic growers do use pesticides, there are residues often on organic products, and whether you consume organic or conventionally grown, you are still going to be consuming a wide array of pesticides which naturally occur in all foods regardless of how they are grown, and in far greater quantities than their synthetic counterparts. She also appears to misunderstand the nutritional value of organics, as well. She further states, "there has been limited research on whether organic foods are nutritionally superior to conventional foods." This is false; see Chapter 4 for more details. As I stated earlier, if the produce is grown under the same conditions, it is quite impossible for there to be any significant differences in nutrient content. The nutrient content is determined by the plant's genetics and the conditions it was grown in, not the method of growth. Cow manure does not enhance the nutrient content of produce.
- On page 116, she states pesticides "may endanger human health and life." There has not been one study illustrating any consumer has ever been harmed by the either zero exposure level on most foods by the time they reach consumers, or the infinitesimally small exposure level when consumers are exposed to them. Again, see Chapter 4 for more details to clarify.
- On page 543, she states, "the FDA [U.S. Food and Drug Administration] estimates that 33 to 39 percent of our food supply contains residues of pesticides that *may pose a long-term danger to our health* [my emphasis]." This is blatantly false. The October 1, 2018 release from the FDA entitled *FDA In Brief: Final Results from FDA's Pesticide Monitoring Program Shows Pesticides Residues in Foods Below Federal Limits* quotes FDA Commissioner Scott Gottlieb, M.D. He states, "today we're releasing the latest set of results from our annual Pesticide Monitoring Program. Like other recent reports, the results show that overall levels of pesticide chemical residues are below the Environmental Protection Agency's tolerances, and *therefore don't pose a risk to consumers* [my emphasis]."[32]

Coaches

I have spoken at coaches' clinics three times a year for 25 years. Our local high school district, the largest in the state of California conducts these for each fall, winter, and spring sport seasons for all new incoming part-time or full-time coaching staff. One of the main purposes of these clinics is to illustrate to the coaches the liability of many of the supplement industry's products, and why coaches are never allowed, under any circumstance, to recommend any over-the-counter product except hydration drinks such as Gatorade, Powerade, or similar products. Additionally, it is well documented that most coaches embrace many of the same areas of misinformation their athletes do, addressing which is one of the goals of the clinic: to better educate the coach which is the main contact person for most athletes.

Most coaches simply do not have the appropriate nutrition or biochemistry science background to be considered a reliable source. This does not mean that some coaches are not well versed in nutrition science, which many are, but they are simply unlikely to be beyond the basics. As an example, in 2012 the *Journal of Athletic Training* published "Sports Nutrition Knowledge Among Collegiate Athletes, Coaches, Athletic Trainers, and Strength and Conditioning Specialists." The researcher's results indicated that only 35.9% of the 131 NCAA Division I, II, and III coaches who participated had adequate knowledge.[33]

Certified Athletic Trainers and Certified Strength and Conditioning Coaches

Both sources can be a reliable source of information for athletes, with some caveats. Not all certified athletic trainers (ATCs) or certified strength and conditioning coaches may come from a strong academic background or have pursued appropriate additional training. As an illustration, in the September/October 2008 National Strength and Conditioning Association's Performance Training Journal, a NSCA-registered strength and conditioning coach, published "Do Athletes Maintaining Healthy, Well-Balanced Diets Really Need Nutritional Supplements."[34] at the time, he was the strength coach for the Major League Baseball Toronto Blue Jays as well as on the board of the Professional Baseball Strength & Conditioning Coaches Society. I use this paper towards the end of my course as a homework assignment. At this time, the students have been exposed to enough information to review this paper and most are able to pick out the 15-plus basic nutrition science errors contained in it. The misinformation in this paper highlights that even at the highest level of professional sports, nonsense and misinformation persists, even from the "professional" staff. This surprises all students, but also clearly illustrates the need for this book.

Here are some examples of the errors in this paper, which can be identified by freshman college students at the end of one semester of a beginning nutrition science course:

- "Organically grown, antioxidant-rich fruits and vegetables represent one of the only nutritious food sources readily available. Unfortunately, getting your well-balanced daily servings of nutrient -rich food is a difficult task (page 10)."

Unless the individuals he has intended this article for live in some Third World country, obtaining high-quality food, especially for professional athletes, is an easy task unless foolish decisions are made by the individual.

- He feels that unless food is organically grown, it is likely "potentially full of toxins due to overuse of fertilizers, pesticides, and herbicides (page 10)." Again, see Chapter 4 for more details as to why he is mistaken here.
- "And, as an athlete, to ensure peak performance and to ward off chronic degenerative diseases, you need to appropriately supplement a well-balanced, whole food-based diet for optimal nutrient (page 10)." This is a contradiction. If the diet is well-balanced, why do you need a supplement? So, even if you make wise food choices, you are going to need a supplement. See Chapter 7 of this book.
- "By design, supplements should supplement, filling in the nutritional gaps of a quality diet. Generally, these nutritional gaps are seen as a lack of vitamins, minerals, and antioxidants (page 10)." Again, as before, contradictory for the same reasons.
- "When examining nutritional deficiencies you need to be aware of the void between the Recommended Dietary/Daily Allowance (RDA) and Optimal Levels of nutrient intake of vitamins, minerals, and antioxidants. The RDA suggests standard intake levels to meet the minimum nutrient requirements for the majority of healthy individuals (page 10)." He has this backwards. The RDAs meet the maximum nutrient needs for most of the population, not the minimum. As an example. The RDA for vitamin C is 90mg per day, which by many researchers estimates, is far more than is necessary to maintain the maximum 1,500mg reserve if it. Vitamin C above 100mg per day will begin to have a significant excretion rate through the kidneys. Additionally, due to the three mechanisms of action discussed in Chapter 7, it is not even necessary to meet the RDAs on a daily basis as discussed. Just come close. This applies to athletes, as well.
- "In some cases, optimal nutrient levels can be as much as thirty times greater than the RDA (page 10)." This is dangerous advice with no scientific basis for it. This advice could possibly exceed the upper limit of safety for some nutrients—the dose makes the poison, not the compound.
- "However, to meet the optimal nutrient levels in some cases would require eating a substantial amount of foods rich in the needed nutrients. This is just not possible in most situations. Consequently, adding a high-quality multivitamin and mineral complex to your diet can ensure optimal levels of nutrients for recovery from intense activities and to boost your body's immune system (page 10)." As nutrient needs increase, the absorption rate of them increases as well as does the recycling of them. He appears to have no concept of this. Again, any industrialized country has more than enough high-quality food available, just use a little personal responsibility and make better food choices. It is not the absence of it, but the ingestion of it, which is the problem for

most. Finally, an athlete's immune system is already optimal as a result of his/her athletic pursuits. Physical training automatically enhances one's immune system. Fatigue and overtraining are generally the serious athlete's biggest problem when it comes to the immunity issues, not nutrient deficiencies, which can only be fixed by rest and appropriate adjustments in training load.

- "Athletes need foods that will sustain their energy over long periods, not create roller coasters of blood sugar highs and lows. As such, you should always try to avoid high glycemic carbohydrates, regardless of the time of day (page 10)." There is absolutely nothing wrong with high glycemic index foods, especially post-training or conditioning, to enhance the delivery of both sugars and amino acids into the muscle as a result of the insulin response to the high glycemic index foods. He even states later in the same paragraph to avoid potatoes, even though many of the athletes who play professional baseball come from South American countries, where one of their staple food items would have been his dreaded potato. I can hear the athlete now who has read his bad advice when he makes the phone call home. "Ma, I grew up on potatoes and became an excellent athlete on potatoes, but now they tell me not to eat potatoes."

- "Despite research and supporting evidence that supplementing a healthy diet is necessary to attain optimal nutrient levels for sports performance (page 11)." It is just the opposite. Read the rest of this book. He is contradictory here once again. If it is a healthy diet, then supplements are a total waste of time and may inhibit development. Read Chapter 8.

These missteps are not necessarily an anomaly among strength coaches at the professional level. Consider the following examples from articles posted on the Professional Baseball Strength and Conditioning Coaches Society's website by individuals who have MS, RD, CSSD, as well as CSCS qualifications.

- "Fat-soluble vitamins are stored in the body whereas water-soluble vitamins are not. Therefore, it is important to not take too high of a dose of fat-soluble vitamins as there could be risk of toxicity and take water-soluble vitamins frequently since they are not stored and are excreted in the urine."[35] Water-soluble vitamins do have a reserve capacity, and they do not have to be taken frequently. See Chapter 7 for details.

- "Boost your immune system. Choose foods that are high in antioxidants such as fruits and vegetables to help keep your immune system healthy and reduce the amount of free radicals that your body builds up during high intensity training. Choose more colorful fruits and vegetables such as blueberries, strawberries, kiwis, oranges, broccoli, carrots and sweet potatoes."[36] Food advice is accurate. The free radical fear is not. Under most circumstances for athlete's, free radicals play a significant signaling role for muscle tissue to adapt to the training load. Free radicals also play a major role in maintaining a healthy immune system. The phagocytic cells, the first responders to any viral or bacterial exposure, use free radical to suppress the foreign molecules. So, instilling free radical fear in athletes is unwarranted. Their immune system is automatically enhanced with training, unless the athlete is overtraining. See Chapter 8 for more details.

Finally, in 2015, according to the Professional Baseball Strength and Conditioning Coaches Society's October 30, 2015 article "Los Angeles Dodgers Nutrition Program,"[37] the Dodgers "have gone 'all in' on eating organic foods throughout the organization." The Dodgers player development staff apparently embraced the same common misunderstandings regarding the purported health and safety benefits of organic foods. See Chapter 4 for more details.

As illustrated, not all strength and conditioning coaches are well prepared to give the appropriate advice. However, as demonstrated in the research mentioned previously, which appeared in the *Journal of Athletic Training*, 77.8% of certified athletic trainers and 81.6% of certified strength and conditioning coaches are prepared to provide sound advice.[38] Locally, I have a certified athletic trainer who can readily fill in my lecture time at the coach's clinics if I am not available to do so.

The biggest advantage an athlete has if they can locate a well-versed ATC or certified strength and conditioning coach is their ability to handle the athlete's complete needs, which can simplify the training and developmental process.

Registered Dietitians

Generally, registered dietitians are an excellent resource for athletes and active consumers, but as with all professionals, there are certainly levels of competence depending upon training. As an example, years ago I was attending a departmental meeting at the junior college where I teach a basic nutrition course in the evenings as an adjunct professor. Sitting next to me was one of the instructors who was a registered dietitian. During the conversation, she stated that she was not that comfortable with the sports nutrition section of the course, and would normally ask a friend of hers, who manages a local GNC, to discuss this issue for her on the required evening. Obviously, I was just a tad appalled that she thought this was a good idea (see Chapter 9 as to why). This is analogous to asking the fox to guard the chicken coop. Even though I offered to take that lecture for her on those evenings, as well as provide enough information where she would feel comfortable lecturing on the topic herself to that population group, she declined.

Now the flip side of this coin are registered dietitians who have pursued the Board-Certified Specialist in Sports Dietetics (CSSD) through the Academy of Nutrition and Dietetics. This population group possess all the academic credentials in biochemistry, physiology, and metabolism to make them an excellent resource.

However, as pointed out in the certified strength and conditioning coaches' section above, there are certainly those who possess the CSSD certification I would disagree with.

Exercise Physiologists

Back in the 1970s, I was initially an agribusiness major at California Polytechnic University in San Luis Obispo California. This changed as my interests changed, which is typical of college students. I happened to take a basic nutrition course as an elective and found the biochemistry material interesting but had no interest in what was available at the time as an academic pursuit for that subject material, dietetics. Pursuing the matter, a little further, I was made aware of an option within the California State University system called the Special Majors Program. This program was designed to allow students who wished to pursue nontraditional academic degrees, not provided by any specific academic department, and combine segments of two different academic areas into one with an advisor from both departments. In my case, this involved combining the appropriate course work from within the exercise physiology option of the kinesiology department with the appropriate biochemistry, advanced nutrient metabolism, etc., from the nutrition department at Long Beach State University. This allowed for a better grasp, I felt, of both the fundamental aspects of the physiological adaptations to work and exercise, as it related to the cardiorespiratory and musculoskeletal systems, as well as a much broader understanding of energy and nutrient metabolism than would have obtained from the exercise physiology or nutrition coursework alone. This, as well as some additional training from UCLA's Drs. Karlman Wasserman and Brian Whipp through their Practicum: Cardiopulmonary Exercise Testing and Interpretation, prepared me for seven postgraduate years running a full exercise physiology lab for a large medical practice. The Practicum was inaugurated in 1982, a year after I graduated from graduate school, by Drs. Wasserman and Whipp at UCLA in response to requests for practical instruction in cardiopulmonary exercise testing. As stated in the current description of this practicum from the David Geffen School of Medicine, The Rehabilitation Clinical Trials Center,

> the program was designed to increase the participants understanding the physiologic basis of gas exchange responses to exercise and to be able to use variables and parameters of clinical exercise tests to meaningfully characterize exercise function. The course is intended for physicians in clinical practice or academics, exercise scientists and laboratory personnel involved in cardiopulmonary exercise testing.[39]

During this time, due to my background in both the exercise physiology and nutrition areas, I was asked by William Jarvis Ph.D., the president of the National Council Against Health Fraud (NCAHF) at the time, to coordinate the NCAHF task force on ergogenic aids, which I did until 2011 when the NCAHF dissolved after the passing of Dr. Jarvis. NCAHF was a private, nonprofit, voluntary health agency that focused upon health misinformation, fraud, and quackery as public health problems. Its positions were based upon the principles of science that underlie consumer protection law. It advocates: (a) adequate disclosure in labeling and other warranties to enable consumers to make truly informed choices; (b) premarketing proof of safety and effectiveness for products and services claimed to prevent, alleviate, or cure any health problem; and (c) accountability for those who violate the law. This site, which belongs to former long-time NCAHF board member Stephen Barrett, M.D., archives many of NCAHF's documents. Dr. Barrett heads Quackwatch.org., where I am a Nutrition and Food Science Advisor.

I state all of this for the following reason. Now, Long Beach State University as well as other college departments in kinesiology, such as Montana State University, have established degree options in exercise physiology and nutrition. An exercise physiologist with a strong understanding of biochemistry, physiology, and nutrient metabolism are also an excellent resource.

The Internet

Avoid the internet like the plague unless you have a degree in the field and know how to navigate through all the nonsense. The internet can be a great tool when used properly, but it can also be far too easy to get fleeced by all the counterfeit science which may appear truthful to the average consumer. There is an endless parade of charlatans ready to hustle any naïve or poorly informed individual. For most consumers, the governing bodies of many athletic and professional organizations can provide good peer-reviewed nutrition science information.

Some additional recommendations are as follows. This is not an exhaustive list:

- American Council on Science and Health (www.acsh.org).
- American College of Sports Medicine (www.acsm.org).
- American Dietetic Association (www.eatright.org).
- Academy of Nutrition and Dietetics (www.scanpg.org).
- American Society for Clinical Nutrition (www.faseb.org/ascn).
- Gatorade Sports Science Institute (www.gssiweb.com).
- Quackwatch (www.quackwatch.org).
- Produce Marketing Association (www.aboutproduce.com).
- SafeFruitandVeggies (www.SafeFruitandVeggies.com).
- Alliance for Food and Farming (www.foodandfarming.org).

Notes

1. *The Cambridge Business English Dictionary*, By Cambridge University Press, p. 788.
2. J. Weissman, M. Magnus, T. Niyonsenga, and A.R. Sattlethight, Sports nutrition knowledge and practices of personal trainers. *Journal of Community Medicine and Health Education* (2013), Vol. 3, p. 254. doi: 10.4172/2161-0711.1000254

3. https://www.consumerreports.org/fruits-vegetables/vegetables-that-are-healthier-cooked/
4. S. Storcksdieck genannt Bonsmann, T. Walczyk, S. Renggli, and R.F. Hurrell, Oxalic acid does not influence nonheme iron absorption in humans: a comparison of kale and spinach meals. *European Journal of Clinical Nutrition* (2008), Vol. 62, pp. 336–41.
5. https://www.consumerreports.org/sugar-sweeteners/low-calorie-sweeteners-may-contribute-to-weight-gain/
6. https://www.consumerreports.org/soda/mounting-evidence-against-diet-sodas/
7. https://www.berkeleywellness.com/healthy-eating/nutrition/article/raw-food-diet-craze
8. *The TB12 Method: How to Achieve a Lifetime of Sustained Peak Performance*, by Tom Brady, p. 218.
9. Ibid, p. 223.
10. Ibid, p. 220.
11. Ibid, p. 223.
12. tb12sports.com/shop/products/tb12-electrolyte-and-protein-subscription-bundle
13. *The TB12 Method: How to Achieve a Lifetime of Sustained Peak Performance*. September 19, 2017, by Tom Brady, p. 223.
14. Ibid, p. 228.
15. Ibid, P. 228.
16. Ibid, p. 228.
17. Ibid, p. 229.
18. https://sciencebasedmedicine.org/dr-oz-and-the-terrible-horrible-no-good-very-bad-day/
19. Ibid, p. 1.
20. Ibid, p.1.
21. Ibid, p.1.
22. Food Can Fix It, by Dr. Oz, p. 29.
23. Ibid, p. 39.
24. Ibid, p. 39.
25. James DiNicolantonio J. et al, L-carnitine in the secondary prevention of cardiovascular disease: Systematic review and meta-analysis. *Mayo Clinic Proceedings*, April 15, 2013, Vol. 88, No. 6, pp. 544–51.
26. *Food Can Fix It*, by Dr. Oz, p. 52.
27. Ibid, p. 278.
28. Ibid, p. 277.
29. https://sciencebasedmedicine.org/detox-what-they-dont-want-you-to-know/
30. Christina Korownyk, et al. *Televised medical talk shows—What they recommend and the evidence to support their recommendations: a prospective observational study*. BMJ (2014), p. 349: g7346.
31. *An Invitation to Health, Your Life, Your Future*, by Diane Hales. Cengage Learning. 18th edition.
32. https://www.fda.gov/news-events/fda-brief/fda-brief-final-results-fdas-pesticide-monitoring-report-shows-pesticide-residues-foods-below
33. T.M. Torres-McGehee, K.L. Pritchett, D. Zippel, D.M. Minton, A. Cellamare, and M. Sibilia, Sports nutrition knowledge among collegiate athletes, coaches, athletic trainers, and strength and conditioning specialists. *Journal of Athletic Training* (2012), Vol. 47, No. 2, pp. 205–11.
34. Donovan Santos, CSCS. Do Athletes Maintaining healthy, well-balanced diets really need nutritional supplements? *NSCA Performance Training Journal* (2008), No. 7.5, p. 10–12.
35. http://baseballstrength.org/minerals-by-rob-skinner-ms-rd-cssd-cscs/
36. http://baseballstrength.org/athlete-eating-guidelines-by-rob-skinner-ms-rd-cssd-cscs/

37. http://baseballstrength.org/los-angeles-dodgers-nutrition-program/
38. T.M. Torres-McGehee, K.L. Pritchett, D. Zippel, D.M. Minton, A. Cellamare, and M. Sibilia, Sports nutrition knowledge among collegiate athletes, coaches, athletic trainers, and strength and conditioning specialists. *Journal of Athletic Training* (2012), Vol. 47, No. 2, pp. 205–11.
39. labs.dgsom.ucla.edu/rossiter/pages/exercise_practicum, p.1.

2

THE "HEALTH NEWS"

Why the Media Are Unreliable

A 2018 Gallup/Knight Foundation survey found that 62% of Americans feel the news is biased and 44% feel the news is inaccurate.[1] However, this survey does not specifically identify to what extent the general public perceives science news to be inaccurate. But, one can readily assume that if the media has difficulty accurately reporting observable and readily understandable events, then the reporting of more complex issues related to science would be even more difficult. To fully understand the magnitude of the problem for science reporting, and why caution is advised prior to embracing the face value of any media report of health/science "news," consider the following comments from well-respected medical journals, a top consumer science education site, a very well-respected researcher from Stanford University Prevention Research Center and Meta-Research Innovation Center (METRICS), as well as others. The following comments illustrate why most consumers understandably find nutrition science (as well as science news in general) contradictory, confusing, and unreliable:

1. August 23, 2018, John Ioannidis, M.D., D.Sc., professor of disease prevention at Stanford University, published the following comment in the Op-ed "The Challenge of Reforming Nutritional Epidemiological Research" in the *Journal of the American Medical Association* (*JAMA*): "Some nutrition scientists and much of the public often consider epidemiologic associations of nutritional factors to represent causal effects that can inform public health policy and guidelines. However, the emerging picture of nutritional epidemiology is difficult to reconcile with good scientific principles. **The field needs radical reform**" [my emphasis].[2]
2. August 24, 2018. Alex Berezow, Ph.D. (microbiology), and Senior Fellow of Biomedical Science at the American Council on Science and Health

(ACSH), published the following in his commentary "John Ioannidis Aims
His Bazooka At Nutrition Science":"Here at ACSH, we have been saying for
a long time that nutrition research is shoddy and mostly wrong."[3]

3. December 20, 2018. HealthNewsReview.org released "A final HealthNews-
    Review.org report card from 3,200+ systematic reviews of health care news
    stories & PR releases." This report stated, "News releases on nutrition, nutri-
    tional supplements or 'natural' interventions tended to be consistently more
    promotional than evidence-based—providing little or questionable data."[4]

4. December 1, 2017. *The Lown Institute*'s commentary "The Epidemic of Junk
    Science" stated:

> As much as half of the biomedical research we fund is infected with reckless
> practices and bias.
>
> Cutthroat academic competition, a headlong rush to publish in "high-
> impact" journals, and scarce funding all lead researchers to cut corners, and
> the self-correcting mechanisms of science can't keep up.
>
> This pattern of careless scientific practices has created a crisis of "irre-
> producibility"—few results of these studies can be reproduced, which calls
> their accuracy into question. And when more studies are conducted based
> on the results of flawed studies, it creates the data equivalent of a house of
> cards.
>
> The driving force behind all this? Money. Hyping positive research find-
> ings, whether or not the studies are well done, is "all in the name of gin-
> ning up more money for research and more prestige for the institution,"
> says Shannon Brownlee, vice president of Lown Institute. And it's a vicious
> circle, because peer reviewers are more likely to recommend a study for
> publication when the author is affiliated with a prestigious institution.[5]

5. October 4, 2016. Ruth Kava Ph.D., Senior Nutrition Fellow with the ACSH,
    in *Reliability of Nutrition Research* questioned:"But now, I am forced to admit
    that much of what we think we know about nutrition probably isn't so."[6]

6. April 11, 2015. Richard Horton M.D., editor-in-chief of *The Lancet*, wrote
    the following after attending a symposium on the reproducibility and reli-
    ability of biomedical research. Horton wrote:

> A lot of what is published is incorrect. I am not allowed to say who made
> this remark because we were asked to observe Chatham House rules. We
> were also asked not to take photographs of slides. Why the paranoid con-
> cern for secrecy and non-attribution? Because this symposium—on the
> reproducibility and reliability of biomedical research—touched on one of
> the most sensitive issues in science today: the idea that something has gone
> fundamentally wrong with one of our greatest human creations. The case
> against science is straightforward: much of the scientific literature, perhaps

half, may simply be untrue. Afflicted with studies of small sample sizes, tiny effects, invalid exploratory analysis, and flagrant conflict of interest, together with an obsession for pursuing fashionable trends of dubious importance, science has taken a turn toward darkness. As one participant put it, "poor methods get results."[7]

7. December 17, 2014. *The British Medical Journal* completed a study entitled "Televised Medical Talk Shows—What They Recommend and the Evidence to Support Their Recommendations." The objective of the study was "to determine the quality of health recommendations and claims made on popular medical talk shows," which were obtained from 40 randomly selected episodes which aired daily in 2013 for *The Dr. Oz Show* and *The Doctors*. The conclusions of the study found the following: "Approximately half of the recommendations have either no evidence or are contradicted by the best available evidence. Potential conflicts of interest are rarely addressed. The public should be skeptical about recommendations made on medical talk shows."[8]

8. November 9, 2009. Marcia Angell M.D., past editor-in-chief of *The New England Journal of Medicine* (*NEJM*), wrote: "It is simply no longer possible to believe much of the clinical research that is published, or to rely on the judgment of trusted physicians or authoritative medical guidelines. I take no pleasure in this conclusion, which I reached slowly and reluctantly over my two decades as an editor of the NEJM."[9]

9. August 30, 2005. John Ioannidis M.D., D.Sc., again, in the journal *PLOS/Medicine* "Why Most Published Research Findings Are False," stated the following: "There is increasing concern that most current published research findings are false. . . . Simulations show that for most study designs and settings, it is more likely for a research claim to be false than true. Moreover, for many current scientific fields, claimed research findings may often be simply accurate measures of the prevailing bias."[10]

Now, if these professional opinions illustrate this much concern for the accuracy and reliability of peer-reviewed papers which appear in professional science journals, then just how inaccurate is the information produced by the general media who must rely on this quagmire of misinformation for their reporting? Can a journalist with little to no science background reporting for a newspaper, a "health" magazine, or a televised news station accurately assess the reliability of what he or she is reporting on?

Let me provide an analogy of using counterfeit money to illustrate how difficult it is for most journalists to convey accurate science information. Last year, in the physical rehabilitation clinic where I work during the day, a patient who owns a local pizza parlor brought in a $50 bill he received from one of his customers. My casual appraisal of it was that the bill was authentic, and I would be happy to take it off his hands. He then pointed out that the bill was counterfeit. He illustrated the minor discrepancies between the counterfeit bill and an authentic $50 bill he also possessed, which I did not initially identify because I had not been trained to perceive these differences. The discrepancies between the two bills were in the details. Once the details were pointed out, it was clear that I would have accepted the counterfeit bill believing it to be authentic when it was worthless. I was embracing an imitation, which had no real value. My faith was in a forged bill until it was demonstrated why the counterfeit bill should be rejected.

Counterfeit science is like counterfeit money. To the untrained eye it looks real, but when you examine the details, it is simply fabricated, taken out of context, or extrapolated beyond what the evidence indicates. It is worthless. The details of counterfeit science do not support the conclusions that are being drawn from the evidence. The evidence indicates that the counterfeit science must be rejected, and this is a major problem for the media. The media too often fails to identify counterfeit science and report it as if it were truth and far too many consumers are far too willing to embrace the health "news" and apply it to their lifestyle.

Following are seven prominent reasons why the health news is unreliable. A discussion of each will follow.

1. Most journalists rely on institutional press releases, which often overstates the study's findings, instead of examining the actual published study themselves.
2. Journalists fail to report the study was conducted on rodents and fail to recognize *the Principle of Toxicology—the dose makes the poison*, not the chemical.
3. Journalists most often have no significant science background, making it difficult for them to separate counterfeit or junk science from real science.
4. The media report is based upon an observational study, which cannot demonstrate a cause-and-effect relationship between two things—only an association.
5. The media fail to recognize that a single study never indicates anything. It must be replicated.
6. The media often fail to recognize that the population group used in the study are not large enough or diverse enough to be reliable.
7. The media fail to recognize the scary carcinogenic conclusions of a "study" have no plausible mode of action.

1. Most journalists rely on institutional press releases, which often overstates the study's findings, instead of examining the actual published study themselves.

As an example, in 2015, The UK peer-reviewed journal *PLOS/One* published the findings of researchers from the College of Medicine, Swansea University in Wales, entitled, "When Medical News Comes from Press Releases—A Case Study of Pancreatic Cancer and Processed Meat."[11] This study was an expanded version of a commentary I published in the *British Journal of Cancer* (BJC) in 2012, which illustrated that the widespread media reporting of this study worldwide was greatly exaggerated because of the media's failure to read the study itself, instead relying on the press release.

The researchers from Wales expanded my commentary, which had provided only three examples of media misreporting, to determine just how many media reports worldwide failed to accurately report on the BJC paper. The researchers identified 312 news reports worldwide. Of those 312 reports, "85.6% of all the stories were derived wholly or largely from a secondary source."[12] In other words, 85.6% of the reporters either failed to evaluate the original data themselves, or were incapable of it. The researchers also found that the shortcomings of the paper raised in my commentary were covered in less than a third of the 312 reports.

If reporters had taken the time to evaluate the actual study, instead of just regurgitating each other's stories, or relying on press releases, they would have read statements in the body of the paper made by the authors themselves, which would have clearly identified the study as counterfeit science and not worth reporting. As an example, the authors clearly declare in the body of the paper the following statements which should have preempted this "study" from ever making it into any media headline:

- **"Our study has some limitations** [my emphasis]. First, as a meta-analysis of observational studies, we cannot rule out that individual studies may have failed to control for potential confounders, which may introduce bias in an unpredictable direction." In other words, they are questioning the reliability of the data they are evaluating.[13]
- "Only a few studies adjusted for other potential confounders such as body mass index and history of diabetes." In other words, they are stating that other lifestyle and personal health issues or variables were not accounted for, which are far more likely to have accounted for the purported increased rate of cancer.[14]
- "Our findings were likely to be affected by imprecise measurement of red and processed meat consumption and potential confounders." In other words, they really had no idea of the personal consumption levels of red and processed meats of the subjects studied.[15]

Any journalist making any credible effort should have picked up on these self-admitted shortcomings and immediately passed on reporting it. Instead, the media worldwide published exaggerated headlines, and instilled unnecessary fear in consumers regarding the purported dangers of processed meats. Here are a few of examples of the misconstrued headlines:

- **CBS News:** "Pancreatic Cancer Risk Increases With Every 2 Strips of Bacon You Eat" (January 13, 2012).

- **Fox News:** "Two Slices of Bacon a Day Increases Cancer Risk by a Fifth" (January 13, 2012).
- **USA Today:** "Study Links Processed Meat With Increased Risk of Pancreatic Cancer" (January 13, 2012).
- **CBN:** "Processed Meat Increases Risk of Pancreatic Cancer" (January 20, 2012).
- **Huffington Post:** "Processed Meat Could Raise Pancreatic Cancer Risk" (January 13, 2012).
- **ABC News:** "A Link Between Sausage and Cancer?" (January 13, 2012).

You can read my original *BJC* commentary if you wish (Lightsey D. Comment on "Red and Processed Meat Consumption and Risk of Pancreatic Cancer: Meta-analysis of Prospective Studies." British Journal of Cancer. 2012;107: 754–755).

When my freshman and sophomore college students who have no significant science background are shown the same information, which was readily available to the media, they are always startled that the media can be so easily duped by such blatantly obvious misinformation. Additionally, once this or any junk science news starts obtaining traction, news agencies around the world, like witless sheep, as illustrated by the study in the journal *PLOS/One*, just begin regurgitating the same misinformation reported elsewhere, without taking the time to clarify if the original reports were accurate and based upon sound science. This is a far-too-frequent problem with the media.

The facts regarding the safety of nitrites in processed meats which the media clearly overlooked are presented in Chapter 3.

2. **Journalists fail to report the study was conducted on rodents and fail to recognize** *the Principle of Toxicology—the dose makes the poison,* **not the chemical.**

Rodent studies can be helpful, and provide possible insight, but they cannot be directly applied to humans due to often vast differences in metabolism as well as general physiology. Rodents are not little people, and you cannot directly correlate what happens in animal studies to humans, yet the media repeatedly do so. Additionally, the doses given to rodents in most studies, greatly exceed those levels humans are exposed to, sometimes hundreds to thousands of times greater than humanly possible, which leads to the next media misunderstanding: *The Principle of Toxicology—the dose makes the poison.* No chemical is inherently bad for you until your exposure level to it exceeds its upper limit of safety. This applies to all chemicals, including vitamins and minerals.

As an example, each year the media loves to publish the Environmental Working Group (EWG) Shoppers Guide "Dirty Dozen" list of the purported top 12 conventionally grown fruits and vegetables consumers should consider buying organic rather than conventionally grown. EWG is heavily funded by the organic food industry, so it will be clear from the following information that they are more agenda driven vs. consumer education driven. The EWG's methodology for determining this ranking lacks scientific credibility, and the media, if they properly researched their stories, should have been aware of this. The media should have recognized years ago that the EWG "Dirty Dozen" list is essentially contrived, and as a result, discontinued their reporting.

On May 15, 2011, the *Journal of Toxicology* published an article by Carl Winter, Ph.D., of the Food Science and Technology Department of the University of California, Davis entitled, "Dietary Exposure to Pesticide Residues From Commodities Alleged to Contain the Highest Contamination Levels."[16] Dr. Winter researches the detection of pesticides and naturally occurring toxins in foods, and how to assess their risks. Dr. Winter, when referring to the EWG methodology for determining risk, states "the methodology used by the environmental advocacy group to rank commodities with respect to pesticide risks lacks scientific credibility."[17] He also states that his department research findings "do not indicate that substituting organic forms of the 'Dirty Dozen' commodities for conventional forms will lead to any measurable consumer health benefit."[18] This is explained in greater detail in Chapter 4. Additionally, Robert Krieger, Ph.D., a Fellow in The Academy of Toxicology Sciences, obtained his Ph.D. from Cornell University where he studied pesticide science, biochemistry, and physiology. Dr. Krieger produced "Perspective on Pesticide Residues in Fruits and Vegetables" when he was Director of the Personal Chemical Exposure Program, Department of Entomology, University of California, Riverside.[19] He maintains an active research program concerning the fate of chemicals, particularly pesticides, in plants, animals, and people. He makes the following comments concerning the EWG Shoppers Guide "Dirty Dozen" list:

- The claim that the Shoppers' Guide shows the fruits and vegetables with the most and least pesticides is erroneous and not supported by published EWG Methodology or personal correspondence with staffers.[20]
- Shoppers are urged to take a careful look at the EWG classification scheme. It is determined by the number of residues (not amount) occurring in produce in the USDA Pesticide Data Program samples. EWG and **uncritical media** [my emphasis] transform the EWG numbers into a notion of potential consumer exposure. For the residues that occurred in the highest amounts of all in the USDA's Pesticide Data Program data from 2000–2008, hundreds to thousands of servings of fruits or vegetables in a single day are required of children, teens and adults to represent a dosage equivalent to the NOAEL! (No Observable Adverse Effect Level). When it comes to exposure, the Shopper's Guide doesn't deliver![21]

- It is groundless to suggest that the Shopper's Guide can be used to meaningfully predict risk. The testing that is used to identify the inherent hazards of pesticides also yields a measure of exposure that is not associated with any detectable adverse effects (toxicity). The pesticide exposures that result from consumption of hundreds to thousands of servings of produce with the very highest residues measured represent no effect levels of exposure.[22]

Let me clarify what all this means to the average consumer. Spinach is normally one of the examples the EWG places on their "Dirty Dozen" list. One of the pesticides (chemicals) used to produce spinach is the insecticide permethrin. It works as a neurotoxin to insects because insects are unable to break it down, whereas humans can, which makes it essentially harmless to humans if they are ever exposed to it. Its effect on insects is why it is a widely used insecticide for agricultural purposes, as well as for mosquito control, head lice, and scabies. The National Pesticide Information Center (NPIC) also points out that less than 1% of the more than 1,700 food samples tested by the USDA had detectable levels of permethrin.[23] This is based upon older USDA data when the NPIC report was published. The most current USDA Pesticide Data Program monitoring for 2017 supports the NPIC and indicated that of the 3,970 food samples tested by the USDA for permethrin, only 67 samples had any detectable residue.[24] This is .016%. Now does this appear to be a real concern for consumers as the EWG portrays it to be? Even if you are exposed to this infinitesimal amount of permethrin, your ability to break it down in the liver detoxifies it, then excretes it, making this minute exposure level a moot point.

But there is more. Permethrin has an established NOAEL of 25mg/kg/day, according to the National Pesticide Information Center.[25] The NOAEL is the exposure level at which no negative health effects can be attributed to the pesticide in question. The EPA uses the NOAEL to calculate the daily safe levels of exposure to any given pesticide. So just how much spinach could an individual safely consume equivalent to the NOAEL dose of permethrin on spinach? According to Dr. Krieger's calculations, this would allow the safe consumption of 3,205 cups of spinach per day for women, 4,487 for men, 3,344 cups for teens 12–19 years old, and 2,564 cups for children 2–5 years old without being harmed by the pesticide.[26] Has any of the annual media reports who publish the "Dirty Dozen" list informed you of this?

Now, here is the real irony in this issue. Acute iron toxicity occurs at roughly 20mg/kg/day, considerably less than the 25mg/kg/day NOAEL set for permethrin. So, if you could theoretically consume the thousands of servings per day of spinach necessary for the pesticide to harm you, if it is on the spinach at all—which is not likely, as indicated by the USDA residue testing program—you would experience toxicity from the naturally occurring iron in the spinach first. But iron is not the only nutrient which would become toxic at those serving levels safe for the pesticide. Vitamin B6 has an established upper end of safety of 200mg per day. Anything over this dose for an extended period of time may result in neurological damage similar to multiple sclerosis. As stated previously, the pesticide for spinach is safe up to 3,200 cups per day, an obvious impossibility as far as consumption is concerned. However, at 3,200 cups per day of spinach, you would be exposed to 704mg of B6, which is over three times the dose necessary to cause neurological damage. Additionally, vitamin C is generally considered safe up to 2,000mg per day, but this value may be much lower. If you consumed the 3,200 cups of spinach, the safe exposure level for the pesticide, you would be exposed to 28,880mg of vitamin C, well over the 2,000mg upper limit of safety for vitamin C. The point here is clear: the "naturally" occurring chemicals in food, even many vitamins and minerals, would be far more toxic to you than the pesticide used to grow the food.

So, is the EWG concern for the pesticide really something to be fearful of? Yet the media, year after year, are exploited by the EWG to publish their junk science and run exaggerated headlines regarding the purported "toxins" on our food. The media are far too vulnerable to agenda-driven vs. science-driven environmental and purported "consumer" groups. But this is all plausible due to the science illiteracy of most journalists.

For more on the credibility problem with the EWG, as well as other anti-consumer and anti-science organizations, the Center for Organizational Research and Education (CORE), provides Activist Facts (www.activistfacts.com). CORE is a 501(c)(3) nonprofit dedicated to research and education about a wide variety of activist groups, exposing their funding, agendas, and tactics.

3. **Journalists most often have no significant science background, making it difficult for them to separate counterfeit or junk science from real science.**

This fact is generally one of the easiest illustrations to use in the classroom to make students aware of the folly of relying on the media for lifestyle advice. Why would anyone rely on someone else with no more science background than themselves, embrace what they hear or read as fact, and then alter their lifestyle habits without validating the information first?

In the college course I teach, I routinely use published studies, as well as the associated media reports of them, to teach students how to more appropriately interpret the findings which are generally far different from how the media reported them. After the 16-week course when students are routinely pressed to evaluate various studies, especially observational studies, most fully grasp the shortcomings of the media, and why they must not depend on the media as a reliable source of science information. It is simply another example of the blind leading the blind. In "Is A Ph.D. Still Useful to Society?" Alex Berezow Ph.D. stated, "most news outlets do not require science writers to have any scientific training at all. (It's not a coincidence, therefore, that most science journalism sucks)."

4. The media report is based upon an observational study, which cannot demonstrate a cause-and-effect relationship between two things—only an association.

Observational study designs are the most commonly reported by the media because they are the most widely used. Nonetheless, observational studies are incapable of demonstrating a cause-and-effect relationship between two things, only an association, even though the media often like to alter their headlines with something far more alarming than the study indicates. Frightening headlines increase readership or viewership, but they are often not based upon the reality of the details of the study.

As an example, I have already mentioned my commentary in the *BJC* in 2012. My commentary was in response to a "study" published in the *BJC* which stated that eating too much processed meat (nitrites) was associated with pancreatic cancer, which resulted in the misleading headline examples provided.

However, the increased pancreatic cancer rates were far more likely attributed to what was missing in the participants' diet vs. the consumed processed meats, which is always the most missed association in diet and disease related studies. Specifically, those who consume large amounts of processed meats are far more likely to consume minimal amounts of fruits or vegetables, which contain literally thousands of plant chemicals likely to play a role in our health. The **absence** of these compounds is far more likely to be the preceptory factor for increased cancer rates than the nitrites in processed meats. There is simply no plausible mechanism of action for the nitrites to be the precipitating factor, whereas there certainly is for the lack of fruits and vegetables in the diet (see Chapter 3 for this reasoning). Additionally, obesity, diabetes, sedentary lifestyle, etc., are just a few of the potentially causative variables which would have to be well controlled. These variables normally are not well controlled for, as was the case in the *BJC* study on processed meats and pancreatic cancer which initiated the misguided, 300-plus international headlines mentioned previously.

Finally, observational studies rely heavily on self-reported lifestyle questionnaires. These are highly inaccurate and are contingent upon the participants reliably and accurately recalling past eating and other lifestyle habits. This type of data collection, especially recall data, is impossible to validate, and the likelihood that most participants are honest regarding their self-reported body weight, exercise duration and frequency, alcohol intake, smoking habits, food intake, etc., is a major obstacle in taking this type of data seriously.

Conducting human, randomized clinical control trials, where the variables are controlled for, can demonstrate or refute a cause-and-effect relationship. However, these types of trials are simply too unfeasible to conduct due to the obvious difficulty in doing so. Consider how impractical it would be for any researcher to control the daily activity, dietary habits, exercise habits, sleep, alcohol intake, smoking, drug use, etc. of a large population group of participants. Therefore, this is why there is so much speculation and flip-flopping in recommendations in the nutrition science area. Much of it is simply defined by speculation and false associations.

5. **The media fail to recognize that a single study never indicates anything. It must be replicated.**

Any study must be replicated many times before it can be considered even a theory, much less a fact. However, the media often report on "new" research as significant prior to serious peer review scrutiny and replication of its findings.

6. **The media often fail to recognize that the population group used in the study are not large enough or diverse enough to be reliable.**

The media often report on studies based upon very small sample sizes. As pointed out in the beginning of this chapter by Richard Horton M.D., editor-in-chief of *The Lancet*, small sample sizes are difficult to extrapolate from.

7. **The media fail to recognize the scary carcinogenic conclusions of a "study" have no plausible mode of action.**

The media will run a frightening headline regarding the purported carcinogenicity of a specific chemical used in the production of our food supply but fail to state that the purported carcinogen is also found in greater quantities naturally in the food itself. Examples of this would be nitrites and aspartame which are covered in the next chapter.

Now this is not an exhaustive list of reasons why the media is unreliable, but it should give the consumer plenty to reflect on when trying to make sense, if possible, of the health news.

Let me provide four media examples when sensational media headlines or statements were unsupported by the actual data being reporting on.

Media Example 1

Headlines:

*   **NBC:** "Highly Processed Foods May Raise Cancer Risk, Study Finds" (February 15, 2018).
*   *USA Today*: "Study Suggests Link Between Consumption of Ultra-Processed Foods and Cancer" (February 15, 2018).
*   **CNN:** "Ultra-Processed Foods Linked to Cancer" (February 28, 2018).

- **BBC:** "Ultra-Processed Foods Linked to Cancer Risk" (February 15, 2018).
- **Daily Mail Online:** "Processed Foods Are Driving Up Rates of Cancer: Major Study Reveals the Health Threat Including Cereal, Energy Bars, Sausages, and Chocolate" (February 14, 2018).

The headlines are based upon the study "Consumption of Ultra-Processed Foods and Cancer Risk," reported in the *British Medical Journal* in February 2018.[26] The study followed the participants on average for five years, collecting what they considered relevant data involving over 100,000 participants' lifestyle habits. These habits included diet, alcohol ingestion, body weight, exercise habits, etc. They obtained the data using an online questionnaire that collected two to three days' worth of daily lifestyle habits every six months.

There are many problems with this study, but here are the five obvious ones— which should have been apparent to any journalists covering this piece and prevented the absurd headlines resulting from it.

First: The study's authors clearly state that the research "assessed the *association* [my emphasis] between ultra-processed food consumption and risk of cancer." As I have already pointed out, an association between two things does not demonstrate a cause-and-effect relationship, just an association.

Second: Self-reported lifestyle questionnaires are highly inaccurate, as I explained earlier. Additionally, the study relied on only a two- to three-day self-reported diet history every six months, which allegedly represented participants' daily dietary habits over the entire six months. This is an unlikely assumption.

Third: The authors attempt to equate much of the increased cancer rates in this population group with the increased intake of various chemicals, which, according to the authors, "have carcinogenic properties." These chemicals include such items as acrylamide, purported endocrine disruptors such as bisphenol A, nitrites, aspartame, etc. The purported carcinogenicity of these compounds is based upon studies of rodents that have been fed volumes of the compounds humanly impossible to consume. As I have already stated, animal studies cannot be directly extrapolated to humans, especially from strains of rodents which may have been used who are predisposed to tumors and would get them regardless of what they were fed. Mice are not little people, nor are we ever exposed to the volume of these compounds given in rodent studies. *The Principle of Toxicology* applies here, and every chemical previously mentioned has well-established safe metabolic pathways in humans.

Fourth: The authors use an unreliable reference as their source for the potential carcinogenicity of the chemicals they highlight in the article: The International Agency for Research on Cancer (IARC). The IARC has gained a reputation of fear-mongering rather than identifying true health risks.

The American Council on Science and Health (ACSH) points out, "IARC has become a fringe group, seemingly more interested in scaring people than identifying actual health threats. Any organization that declares bacon to be as dangerous as plutonium has entirely lost its way" ("Glyphosate-Gate: IARC Scientific Fraud").[27]

The ACSH also points out in an April 20, 2017 report, "The IARC Credibility Gap and How To Close It,"[28] that the IARC program on "Evaluation of Carcinogenic Risks" must be reformed and brought into the 21st century—"or it should be abolished." Additionally,

> the IARC has continued to apply its classification system largely as if the last half-century of scientific research hadn't happened, completely ignoring issues of dose and exposure that are fundamental to risk assessment as it has been practiced around the world for several decades. The result is an unhelpful, even absurdist, scheme, in which chemicals with orders of magnitude differences in cancer potency are placed in the same group.

For an excellent analysis of the IARC junk science peddling, and how the "IARC has made a name for itself not through prestigious research, but by its controversial involvement pushing political agendas and bowing to activist researchers," I suggest the investigative report "Peddlers of Junk Science—Anti-Chemical Activists Use Flimsy Cancer Claims to Tarnish Industries," published in the *Washington Times*, July 24, 2017 by Richard Berman.[29] This investigative report is an example of good media investigative reporting. Mr. Berman's final comments in the article illustrate the political and financial agenda which directs much of the chemophobia. Mr. Berman states,

> When IARC began widely evaluating possible sources of cancer in 1987, it found that most suspected carcinogens, well, weren't. But "toothbrush bristles probably don't cause cancer" can't quite justify a $50 million budget. And as purse strings tighten at the World Health Organization (WHO), IARC's recent reviews continue uncovering a curious number of likely carcinogens.
>
> But the agency's sensationalist tendencies have clearly ventured a step too far. A "WHO insider" confided to Reuters reporter Kate Kelland of talk here now of needing to rein IARC in.

Groups like IARC have the benefit of hiding their politics behind the clout of an international health organization. Unless industries want their products raked over the coals by an agency whose existence depends on the promotion of public fear, they need to join the ranks of leaders speaking out against IARC's tactics.

Fifth: The media and researchers failed to identify the obvious causative variable (poor overall lifestyle and dietary habits, or what I refer to as the LSD diet, the acronym for a Lousy Stinking Diet) and instead focused on specific chemicals they wish to cherry-pick and demonize. An "LSD" is exactly what the participants who were experiencing higher rates of cancer were on. The study clearly points out that those participants whose diets contained up to 50% of their total daily calories consuming ultra-processed foods were experiencing higher rates of cancer. Well, this amounts to a big DUH. The extreme lack of fruits, vegetables and whole grains in the diets of those experiencing higher rates of cancer should have made the conclusion of this study obvious. As an example, consider these compounds from produce and grains that play a role in the prevention of cancer: alkylresorcinols, carotenoids, flavonoids, isoflavones, indoles, isothiocyanates, lignans, monoterpenes, tannins, etc., to name just a few of the thousands of plant chemicals missing from the diets. Additionally, if the participants in question adhered to such abysmal dietary habits, then they are likely to follow just as appalling a routine in relation to other lifestyle habits.

The essential point: The moderate inclusion of processed foods will not increase your risk for cancer; the exclusion of produce and grains will. Adhering to a completely irresponsible lifestyle, clearly indicated by the high-risk participants in this study, will certainly lead to increased rates of a myriad of negative health consequences. This is common sense, and the media should have recognized it.

Media Example 2

Headlines:

- *New York Times*: "Diet Sodas Tied to Dementia & Stroke" (April 20, 2017).
- *Washington Post*: "Study Links Diet Soda to Higher Risk of Stroke, Dementia" (April 21, 2017).
- **NBC**: "Diet Sodas May Raise Risk of Dementia & Stroke, Study Finds" (April 20, 2017).
- **CNN**: "Diet Sodas May Be Tied to Stroke, Dementia" (April 20, 2017).
- *USA Today*: "Diet Soda Can Increase Risk of Dementia and Stroke, Study Finds" (April 21, 2017).
- *Boston Herald*: "Diet Soda Can Increase Risk of Dementia and Stroke, Study Finds" (April 21, 2017).
- *US News and World Report*: Health Buzz: "Drinking Diet Soda Linked to Stroke, Dementia Risk, Study Finds" (April 21, 2017).

These headlines are based on a study which appeared in the journal *Stroke* in April 2017, titled "Sugar and Artificially Sweetened Beverages and the Risks of

Incident of Stroke and Dementia."[30] The authors state, "In conclusion, artificially sweetened soft drink consumption was associated with an increased risk of stroke and dementia."[31]

The discussion on this topic developed in my evening college nutrition course as a result of walking into class with a Wild Cherry Diet Dr Pepper in my hand, and leaving it sitting on the table in front of me. Toward the end of the class, we were discussing the expected higher rates of dementia and strokes in the students' generation later in life due to self-induced high blood pressure from their heavy use of energy drinks and pre-workout drinks (often loaded with various stimulants, both legal and illegal), as well as the increasing rates of obesity.

One of my nursing students politely stated, "But professor, I just read that diet drinks will do the same thing." She of course had zeroed in on my Diet Dr Pepper sitting on the table. I love this type of student for two reasons. One, she is reading outside of the required material for the course, illustrating a real interest in learning, and two, she is bright enough and interested enough that when provided with more of the actual details of the "science" behind a specific topic, she is able to recognize the misinformation.

The student was referring to the recent media reports that originated from a study which appeared in the journal *Stroke* mentioned previously. What basic science principles did the media fail to recognize that should have made a difference in their headlines?

Principle 1: To make a cause-and-effect relationship claim between two things, the study design must lend itself to determining it. The study published in *Stroke* was an observational study, which cannot determine a cause-and-effect relationship between two things, only a correlation, as stated before. In other words, it may be true that the participants who had higher rates of strokes and dementia may have had higher rates of diet soda use, a correlation, but it does not mean that the specific correlation of the diet soda caused the increased rates. Additionally, the actual soda use by any study participant is questionable due to the way the data was collected: using a food frequency questionnaire, which, as stated before, is highly unreliable.

Principle 2: Don't look beyond the obvious; stop the data dredging. The researchers did not control for obesity, a major preceptory factor for both conditions. Additionally, comments from the National Library of Medicine (NLM) report "Behind the Headlines" stated this regarding the study: "Overall, when taking account of all health and lifestyle factors that could have an influence... *there was actually no link between artificially sweetened drinks and risk of dementia* [my emphasis]." The NLM report also stated, "For stroke the links with artificially sweetened drinks were inconsistent. *There were no overall links* [my emphasis] when looking at longer term patterns."

Principle 3: Provide a plausible mechanism of action. There is no plausible mechanism of action that artificial sweeteners could cause stroke or dementia.

One of the most common artificial sweeteners in diet soda is aspartame, which has an extensive research base illustrating its safety. As an example, aspartame is used in my Diet Dr Pepper and is a simple combination of the two amino acids aspartic acid and phenylalanine linked to a methanol, all naturally occurring compounds. All three compounds are separated into individual molecules prior to absorption and have well-established normal metabolic pathways. The body makes aspartic acid, which is also widely dispersed in food in much higher amounts than found in aspartame. Phenylalanine is an essential amino acid the body needs to make all protein molecules, and unless you have phenylketonuria (which is the inability to metabolize it), phenylalanine is metabolized no differently than from any other food in which it is naturally occurring. As with aspartic acid, it is also ingested in far greater concentrations naturally from food.

The methanol occurs naturally in food and is a normal byproduct of digestion. Some will fear the purported dangers of the formaldehyde byproduct of methanol digestion, but every cell in your body produces formaldehyde as part of its normal metabolism. There is also far more formaldehyde produced from the digestion of fruit juice than aspartame. For a detailed review of the safety of sugar substitutes, I recommend the report "Sugar Substitutes & Your Health" from the American Council on Science and Health, or view the brief video on this topic from the American Chemical Society at: www.acs.org/content/acs/en/pressroom/newsreleases/2015/june/is-aspartame-safe-video.html.

Media Example 3

Headlines:

- **WebMD:** "Green Coffee Beans May Aid Weight Loss" (March 28, 2012).
- *Los Angeles Times:* "Green Coffee Beans Show Potential for Losing Weight" (March 27, 2012).
- **ABC News** *Good Morning America:* "Coffee Bean Extract Linked to Weight Loss" (March 27, 2012).
- **CBS News:** "Green Coffee Beans May Lead to Weight Loss, Study Shows" (March 28, 2012).
- *The Dr. Oz Show:* September 10, 2012, Episode "The Fat Burner that Works": "One of the most important discoveries I believe we've made that will help you burn fat—green coffee bean extract."

In 2012, the journal *Diabetes, Metabolic Syndrome and Obesity* published a "randomized, double-blind, placebo-controlled, linear dose, crossover study to evaluate the efficacy and safety of a green coffee bean extract in overweight subjects."[32] The purpose of this study was to determine if a commercially available green tea extract product would produce any significant weight loss in obese subjects and assist in the efforts of preventing obesity. Following is the data available to any journalist who cared to read anything but the press release for this study.

- The study lasted 22 weeks and included 16 overweight individuals.
- According to the paper, "the subjects averaged slightly over an 8 kg weight loss." This is 17.6 pounds over the 22-week time period, which if averaged out, indicates 0.8 pounds of weight loss per day.
- What does this 0.8 pounds mean if it were fat loss? There are 3,500 calories in one pound of body fat. So, this 0.8 pounds per day would mean these subjects somehow experienced a magical 2,800 calories of fat vanishing from their frames per day (0.8 × 3500) "with no significant changes to diet over the course of the study," according to the authors.
- The participants' daily caloric intake, which was not changed for this study, averaged 2,443 calories per day according to Table 3 of the paper.
- In the discussion section of the paper, the authors state that the mechanism of action for this miracle "are unknown." Well, of course not—there is no mechanism of action for magic. It's just magic.
- So, you have individuals who are likely sedentary most of the day, don't change their dietary habits, take a magical green coffee extract, lose more stored calories per day than they are ingesting, and this is supposed to be embraced as "science" worthy of media headlines?

The facts: On January 26, 2015, the Federal Trade Commission (FTC) published "Spilling the Beans: The Anatomy of a Diet Craze."[33] Following is only part of the FTC statements made in this settlement review. Consumers are advised to read the full report.

Some people call it the "Oz Effect"—the bump in consumer demand after a product or ingredient is featured on *The Dr. Oz Show*. In a just-announced settlement, the FTC says defendants Lindsey Duncan, Pure Health LLC, and Genesis Today, Inc., took advantage of that phenomenon by deceptively touting the purported weight loss benefits of green coffee bean extract. The complaint recounts how defendant Duncan fueled a diet craze. The story has it all: a national TV appearance, a popular host, claims that people would lose weight without diet or exercise, and a comprehensive marketing campaign designed to take advantage of all of the above. The only thing missing, according to the FTC, was proof the product actually worked.

What about those striking weight claims? According to the FTC, the defendants didn't have substantiation to support that people would lose 17 pounds in 12 weeks [FTC typo—should say 22 weeks, not 12] and 16% of body fat, without diet or exercise. The complaint also alleges that their claim that a clinical study proved those results was false. **(The** so-called "study" of green coffee bean extract was so hopelessly flawed that the company behind it was the subject of an earlier FTC law enforcement action, resulting in a $3.5 million settlement. In a separate development, the American authors of the study have since retracted it) [my emphasis].

The reader can find the study online and will notice embedded across the front of the paper is the word "retracted."

Media Example 4

Headlines:

- *USA Today*: "Eggs Linked to Higher Risk of Heart Disease and Early Death, Study Says" (March 15, 2019).
- *US News and World Report*: "Eggs Are Bad Again: Study" (March 15, 2019).
- **CNN**: "Three or More Eggs Per Week Increase Your Risk of Heart Disease and Early Death, Study Says" (March 15, 2019).
- *Wall Street Journal*: "Study Links Eggs to Higher Cholesterol and Risk of Heart Disease" (March 15, 2019).

The study was a March 2019 *Journal of the American Medical Association* published paper, "Association of Dietary Cholesterol or Egg Consumption With Incident Cardiovascular Disease and Mortality."[34]

The problems with the study were many, all of which makes any link to increased cardiovascular incidence or disease an impossible connection. Here are a few points as to why this study should not be taken seriously.

- As the study title indicates, this was an association study, so no cause-and-effect relationship can be established, just an association. It is simply cherry-picking one of many variables which could have created the outcome.
- It was a statistical evaluation of a combined six other studies, all of which were performed differently. Statistics cannot demonstrate anything, just a correlation.
- The study relied upon dietary recall questionnaires, which are highly unreliable as stated before.
- Other more likely lifestyle issues which would have accounted for the results, such as fat mass, smoking, exercise habits, sleep, stress, etc., were not properly controlled for, because they cannot be due to the complex nature of all participants' daily lifestyles.

The bottom line: This study essentially cherry-picked a variable, statistically associated it with a disease, and then attempted to demonize eggs and cholesterol with the results. This is a classic example of how junk science is detrimental to everyone. This study may financially impact the egg industry, and it may inhibit many consumers from purchasing a relatively cheap source of quality protein, as well as other nutrients, especially for children.

Recommendation: Don't toss the bacon and eggs. Toss the junk science.

The Take-Home Message for This Chapter

The media can produce good science and health reports as well as good investigating reporting, some of which you will find in this book, but you essentially need a degree in the field they are reporting on to really know if science or junk science is being presented. Unfortunately, the media are not reliable in separating the facts from the hype, and therefore, consumers should be very cautious and selective as to the application of any science or health news coming from the media to their daily lives.

Notes

1. https://news.gallup.com/opinion/gallup/235796/americans-misinformation-bias-inaccuracy-news.aspx
2. J.P.A. Ioannidis, The challenge of reforming nutritional epidemiologic research. *JAMA* (2018), Vol. 320, No. 10, pp. 969–70.
3. www.acsh.org/news/2018/08/24/john-ioannidis-aims-his-bazooka-nutrition-science-13357
4. https://www.healthnewsreview.org/2018/12/a-final-healthnewsreview-org-report-card-from-3200-systematic-reviews-of-health-care-news-stories-pr-releases/
5. https://lowninstitute.org/news/in-the-news/junk-science-is/
6. https://www.acsh.org/news/2016/10/04/reliability-nutrition-research-questioned-10249
7. R. Horton, Offline: What is medicine's 5 sigma? *The Lancet* (April 11, 2015), Vol. 385, No. 9976, p. 1380.
8. C. Korownyk, et al. Televised medical talk shows—what they recommend and the evidence to support their recommendations: a prospective observational study. *BMJ* (2014), Vol. 349, g7346.
9. M. Angell, Drug companies & doctors: A story of corruption. *The New York Review of Books*, January 19, 2009.
10. J.P.A. Ioannidis, Why most published research findings are false. *PLoS Medicine* (2005), Vol. 2, No. 8, p.124. https://doi.org/10.1371/journal.pmed.0020124
11. J.W. Taylor, M.L., Elizabeth Ashley, A. Denning, B. Gout, K. Hansen, T. Huws, L. Jennings, S. Quinn, P. Sarkies, A. Wojtowicz, and P.M. Newton, When medical news comes from press releases—A case study of pancreatic cancer and processed meat. PLoS One (2015), Vol. 10. https://doi.org/10.1371/journal.pone.0127848
12. Ibid, p. 4.
13. S.C. Larsson and A. Wolk, Red and processed meat consumption and risk of pancreatic cancer: Meta-analysis of prospective studies. *British Journal of Cancer* (2012), Vol. 106, No. 3, pp. 603–7. doi:10.1038/bjc.2011.585
14. Ibid, p. 7.
15. Ibid, p. 7.
16. C.K. Winter and J.M. Katz, Dietary exposure to pesticide residues from commodities alleged to contain the highest contamination levels. *Journal of Toxicology* (2011), Vol. 2011, p. 589674.

17. Ibid, p. 1, abstract.
18. Ibid, p. 16.
19. https://faculty.ucr.edu/~krieger/Perspective%20on%20Pesticide%20Report%20 by%20Dr%20%20Krieger.pdf
20. Ibid, p. 4.
21. Ibid, p. 5.
22. Ibid, p. 5.
23. K. Toynton, B. Luukinen, K. Buhl, and D. Stone, Permethirn General Fact Sheet, National Pesticide Information Center (2009), Oregon State University Extension Services. http://npic.orst.edu/factsheets/PermGen.html.http://npic.orst.edu/ factsheets/PermGen.html
24. https://www.ams.usda.gov/sites/default/files/media/2017PDPAnnualSummary.pdf
25. K. Toynton, B. Luukinen, K. Buhl, and D. Stone. Permethirn Technical Fact Sheet, National Pesticide Information Center (2009), Oregon State University Extension Services.
26. T. Fiolet et al., Consumption of ultra-processed foods and cancer risk: results from NutriNet-Santé prospective cohort, *The British Medical Journal* (February 14, 2018), p. 360.
27. https://www.acsh.org/news/2017/10/24/glyphosate-gate-iarcs-scientific-fraud-12014
28. https://www.acsh.org/news/2017/04/20/iarc-credibility-gap-and-how-close-it-11167
29. https://www.washingtontimes.com/news/2017/jul/24/chemical-industries-tarnished-by-junk-science/
30. M. Pase et al., Sugar- and Artificially sweetened beverages and the risks of incident stroke and dementia. *Stroke* (April 20, 2017),Vol. 48, No. 5, pp. 1139–46.
31. Ibid, p. 1146.
32. J.A.Vinson, B.R. Burnham, and M.V. Nagendran, Randomized, double-blind, placebo-controlled, linear dose, crossover study to evaluate the efficacy and safety of a green coffee bean extract in overweight subjects [retracted in: *Diabetes, Metabolic Syndrome and Obesity* (January 18, 2012), pp. 21–7.
33. https://www.ftc.gov/news-events/blogs/business-blog/2015/01/spilling-beans-anatomy-diet-craze
34. V.W. Zhong, L.Van Horn, M.C. Cornelis et al. Associations of dietary cholesterol or egg consumption with incident cardiovascular disease and mortality. *JAMA.* (2019), Vol. 321, No. 11, pp. 1081–95.

3

CHEMOPHOBIA AND THE BOY WHO CRIED WOLF

According to the United Nations Food and Agriculture Organization, in 2019, some 821 million people around the world were undernourished.[1] In contrast, in the United States, as well as most developed countries, over half of the population was either obese or overweight and gluttony was the norm. In the United States, we are a population of people living in an affluent culture whose standard of living is high compared to other nations. Yet we fail to be grateful for the advances in food science and biotechnology we benefit from, which frees us from the day-to-day task of our personal food production. We instead find some new, irrelevant issue to make ourselves miserable with, or complain about, and attempt to demonize just about every advancement food technology has produced to make our lives more comfortable. One of the major phobias many consumers struggle with is related to "chemicals" in the food supply and the many misconceptions about their source, use, safety, and how we all benefit from them. This chapter will illustrate that the fear of chemicals in the food supply is an irrational fear.

To avoid some purportedly dangerous chemicals in food, many consumers have undertaken the growing tendency to embrace the misguided concepts of eating all "natural" or "clean." These are terms which have no basis in science or logic, unless you are referring to clean after washing dirt off your produce. As I point out in Chapter 4, all foods, regardless of how they are grown, contain far more naturally occurring potential toxins and carcinogens (chemicals) than synthetic ones. Terms such as eating "natural" or "clean" are simply marketing terms used to exploit consumer misunderstandings about chemistry, and *the Principle of Toxicology—the dose makes the poison.* This principle, as stated in Chapter 2, is applicable to all molecules, including vitamins, minerals, and the thousands of naturally occurring plant chemicals (phytochemicals), as well as all synthetic compounds. It is irrelevant whether the chemical is either naturally occurring or synthetically produced. Toxicity is related to the exposure level and duration, as well as whether the chemical is stored in the body or broken down and excreted soon after exposure.

As an example, consider caffeine in the cup of coffee you may be drinking while reading this chapter. That naturally occurring chemical is harmless at normal exposure levels, yet, at higher doses, caffeine can cause insomnia, headaches, heart palpitation, tachycardia, jitters, nausea, anxiety, sweating, dizziness, and even cardiac arrest. These symptoms have been extensively documented in the literature, but due to the low exposure level and the fact that caffeine does not accumulate in the body (it is metabolized in 4–6 hours), its potential toxicity is irrelevant to its normal use.

Consider another naturally occurring chemical, cyanide, which is one of the top ten most poisonous chemicals on the planet. The Centers for Disease Control and Prevention (CDC) publication *Facts About Cyanide* points out that cyanide:

> is released from natural substances in some foods and in certain plants such as cassava, lima beans, and almonds. Pits and seeds of common fruits, such as apricots, apples, and peaches, may have substantial amounts of chemicals which are metabolized to cyanide. The edible parts of these plants contain much lower amounts of these chemicals.[2]

Cigarette smoke also contains cyanide. However, the CDC points out that cyanide is metabolized in the liver and then excreted in the urine and small amounts can be converted to carbon dioxide and exhaled. Most importantly, cyanide leaves the body within one day.

The takeaway point here is that yes, both caffeine and cyanide hypothetically can be toxins, but due to your low exposure level to them and your body's ability to metabolize and excrete them, their hypothetical toxicity becomes a moot point. You are simply unable to ingest the dosage levels of either chemical which would actually reach their upper limits of safe exposure levels.

Following is a list provided by the American Council on Science and Health (ACSH) in its "Holiday Dinner Menu" of naturally occurring mutagens and carcinogens found in foods and beverages.[3] This list will further illustrate that if consumers removed or avoided all foods which contained chemicals with the hypothetical potential to be toxins, you would likely have literally nothing left to eat.

- **Acetaldehyde** (apples, bread, coffee, meat, tomatoes): mutagen and potent carcinogen.
- **Acrylamide** (bread, rolls): rodent and human neurotoxin; rodent carcinogen.
- **Aflatoxin** (nuts): mutagen and potent rodent carcinogen; also, a human carcinogen.

- **Allyl Isothiocyanate** (arugula, broccoli, mustard): mutagen and rodent carcinogen.
- **Aniline** (carrots): rodent carcinogen.
- **Benzaldehyde** (apples, coffee, tomatoes): rodent carcinogen.
- **Benzene** (butter, coffee, roast beef): rodent carcinogen.
- **Benzo(A)pyrene** (bread, coffee, pumpkin pie, rolls, tea): mutagen and rodent carcinogen.
- **Benzofuran** (coffee): rodent carcinogen.
- **Benzyl Acetate** (jasmine tea): rodent carcinogen.
- **Caffeic Acid** (apples, carrots, celery, cherry tomatoes, coffee, pears, grapes, lettuce, mangos, potatoes): rodent carcinogen.
- **Catechol** (coffee): rodent carcinogen.
- **Coumarin** (cinnamon in pies): rodent carcinogen.
- **1,2,5,6-Dibenz(A)anthracene** (coffee): rodent carcinogen.
- **Estragole** (apples, basil): rodent carcinogen.
- **Ethyl Alcohol** (bread, red wine, white wine, rolls, tomatoes): rodent and human carcinogen.
- **Ethyl Acrylate** (pineapple): rodent carcinogen.
- **Ethyl Benzene** (coffee): rodent carcinogen.
- **Ethyl Carbamate** (bread, rolls, red wine): mutagen and rodent carcinogen.
- **Furan and Furan Derivatives** (bread, onions, celery, mushrooms, sweet potatoes, rolls, cranberry sauce, coffee): many are mutagens.
- **Furfural** (bread, coffee, nuts, rolls, sweet potatoes): furan derivative and rodent carcinogen.
- **Heterocyclic Amines** (roast beef, turkey): mutagens and rodent carcinogens.
- **Hydrazines** (mushrooms): mutagens and rodent carcinogens.
- **Hydrogen Peroxide** (coffee, tomatoes): mutagen and rodent carcinogen.
- **Hydroquinone** (coffee): rodent carcinogen.
- **D-Limonene** (black pepper, mangos): rodent carcinogen.
- **4-Methylcatechol** (coffee): rodent carcinogen.
- **Methyl Eugenol** (basil, cinnamon, and nutmeg in apple and pumpkin pies): rodent carcinogen.
- **Psoralens** (celery, parsley): mutagens; rodent and human carcinogens.
- **Quercetin Glycosides** (apples, onions, tea, tomatoes): mutagens and rodent carcinogen.
- **Safrole** (nutmeg in apple and pumpkin pies, black pepper): rodent carcinogen.

As one can see, a chemical-free diet is an unnecessary phobia. The synthetic chemicals which have been added to the food supply add a wide variety of safe and necessary functions such as the prevention of spoilage, mold growth, rancidity, antioxidants, emulsifiers, coloring and flavoring agents, nutrients, and many others. This can include any substance used in the production, processing, treatment, storage, packaging, or transportation of the food—as examples, salt to preserve meat, herbs and spices for flavoring, and sugar to preserve fruit. Those who fear these things could eliminate them if they are willing to grow, harvest, grind, can, and cook their own food and then accept the limited choices they would have. Also, be prepared to spend all your "free" time repeating this cycle to feed a family, and hopefully do a sufficient enough job to prevent spoilage as well as food poisoning.

Ironically, consumers who embrace all "natural" and "clean" eating also wisely embrace sustainable agriculture—which is the goal of conventional agriculture, as well. But the two ideologies are incompatible to the other. To be sustainable, the food production process must develop three specific methods of food production for the growing population:

1. One which limits the acreage we need to grow our food on by maximizing the yield per acre, as well as reducing the water needed to do so, which conventional agriculture does.
2. One which, once the food has been produced, provides the technology which allows us to extend the shelf life and availability of the food produced.
3. One which maximizes the safety of the food produced for the growing population, which many synthetic as well as many naturally occurring chemicals do.

So, it is counterintuitive when consumers reject the very technology which allows millions of people to have available to them high-quality food at a cost they can afford through sustainable agricultural practices. Having a disdain for synthetic chemicals, which means you reject modern agricultural practices, means you are willing to embrace primitive food production practices, which is not sustainable, nor will it be affordable for many. This is illogical when we have the technology to ease the food production burden and availability for millions of people.

Unfortunately, due to the relentless onslaught of counterfeit science and the misinterpretation of observational studies, most consumers have developed an unnecessary phobia to synthetic chemicals. Many consumers readily embrace what they read in a purported health magazine, see online, or observe on television as factual vs. a complete misrepresentation of what the actual data illustrates. I demonstrate this in Chapter 2, as well as Chapter 4.

All synthetic chemicals used in the food production and distribution process go through a gauntlet of safety and efficacy studies, unlike the supplements most Americans ingest, which go through zero. Yet, most of these supplements contain synthetic chemical versions of the nutrient. Additionally, just because a chemical is considered "natural" certainly does not mean it is safe. According to James Kennedy, who obtained his master's degree from University of Cambridge in Natural Sciences and is the author of *The Naturalness Fallacy*, **nine out of the top ten most dangerous compounds on Earth are naturally occurring** [my emphasis]. These are botulinum toxin (as illustrated previously), tetanospasmin, palytoxin, diphtheria toxin, ciguatoxin, batrachotoxin, ricin, saxitoxin, and tetrodotoxin.[4]

Even vitamins and minerals (natural chemicals) have established upper limits of safety, which can be found in any basic college level nutrition text. As an example, consider the naturally occurring chemical vitamin B6, which is found in all plant food and is essential to your health. If you exceed vitamin B6 upper limit of safety of 100mg for an extended period, you can develop neurological damage which mirrors multiple sclerosis. Additional symptoms may include nausea, heartburn, and sensitivity to sunlight, as well as unsightly skin patches, according to the National Institutes of Health.[5] This may occur in teenagers as low as 80mg, but these are intakes levels impossible to achieve from the ingestion of food alone, which is one of the main points of this chapter, *the Principle of Toxicology—the dose makes the poison.*

Following are nine examples of various common chemicals in the food supply which have received a significant amount of unnecessary stigma. These examples are provided to illustrate how junk or counterfeit science can stigmatize harmless chemicals we benefit from. Most often, this stigma is due to poorly applied observational studies, which cannot demonstrate a cause-and-effect relationship between two things, only an association, and high-dose rodent studies, which cannot be directly applied to humans. The apparent association to a specific negative health effect of each chemical can be far more attributable to other more relevant issues in the individual's lifestyle habits or physical condition rather than the chemical ingredient. These other relevant issues could include obesity, sedentary lifestyle, chronic poor food choices, excess alcohol, drugs, smoking, etc.

Aspartame

Aspartame is arguably one of the most stigmatized additives used in the food supply process. Its safety is unquestionable based upon its chemical composition and the normal metabolic pathways for those compounds. However, I have already reviewed this ingredient in Chapter 2 under Media Example Number Two. Please refer to that section for the safety of aspartame.

Nitrites

Recently an intern at the physical therapy practice where I work and who is finishing his kinesiology degree at a local university happened to ask me what my take was on nitrites in processed meats. He apparently had just finished a course where the professor had demonized nitrites with the standard misunderstandings of its purported safety concerns.

In processed meats, the nitrites function as an antioxidant, prevent botulism, and maintain color, aroma, and flavor, as well as increase the shelf life of the product. Keep in mind that botulism will quickly remove you from existence.

No forwarding mail address. Botulin botulinum toxin, which is one of the most poisonous biological substances known, is a neurotoxin produced by the bacterium *Clostridium botulinum* and produces several neurotoxins. The neurotoxins block the release of the neurotransmitter acetylcholine, resulting in paralysis of your muscles, including your diaphragm, which makes it difficult to breath and possibly shortens your lifespan to just a few days. According to Cornell University Scientific Inquiry Series—Student Edition publication *Assessing Toxic Risk*, "the compound that causes botulism is a million times more toxic than cyanide, and twenty million times more toxic than caffeine. In its pure form, less than one drop of botulin toxin is enough to kill 500 adult humans."[6] So it should be very evident to the reader that the prevention of this toxin in our food is a wise choice.

Nitrites are a natural product derived from the breakdown of nitrates. Nitrates occur naturally in large quantities in a variety of vegetables, as well as in your saliva, which are the sources for approximately 80–90% of the naturally occurring nitrates you consume. When nitrate from food meets bacteria present in the mouth and intestines, it is reduced to nitrite. This is normal. As an example, just one-fourth of a cup of spinach contains 370mg of naturally occurring levels of nitrates, which the body will naturally convert to roughly 30mg of nitrite at the known 8% conversion rate. This is a naturally occurring process and no one will suggest that spinach is bad for you, except maybe your 5-year-old and some journalists, which unfortunately, they do (see the example in Chapter 2). So, given that spinach would increase my nitrite exposure level greater than processed meats, should I forgo the spinach and bring on the ice cream? These naturally occurring exposure levels far exceeds what you are minimally exposed to in any processed meats such as bacon, hot dogs, salami, etc. If the nitrite in the processed meat is harmful, then spinach is likely going to kill you, as well as green beans, lettuce, and many other vegetables, which is of course baloney—excuse the pun. Enjoy your bacon.

High Fructose Corn Syrup (HFCS)

A good diet without HFCS (corn sugar) also contains significant fructose, as it is one of the principal sugars in most fruits. Depending upon the types of fruit chosen, a normal diet can easily match or surpass the fructose intake of a can of soda. As an example, my diet daily includes considerable raisins, several bananas, apple, etc., which, due to the volume needed to match my energy needs, surpasses the fructose intake I would obtain from one can of soda. HFCS is roughly 55% fructose and 45% glucose, which is similar to table sugar, which contains 50% fructose and 50% glucose. So, there is very little distinguishable difference between HFCS and table sugar (sucrose).

In 2012, the journal *Nutrition and Metabolism* published the study "Fructose Metabolism in Humans—What Isotopic Tracer Studies Tell Us."[7] This study reviewed 19 relevant tracer studies, which illustrated that little dietary fructose appears in circulation due to the chemical conversions that take place in the liver after absorption. The liver converts roughly 50% of consumed fructose to glucose, which is the main energy substrate for the brain as well as other tissues. Another 25–30% of fructose is converted to lactate, a chemical cousin of glucose, which the liver, as well as other cells, can also use as an energy source. Overall, less than 1% of the ingested fructose in humans is converted to fat—the balance metabolically ends up as glycogen (stored sugar), glycerol and carbon dioxide (CO_2).

In 2018, in the journal *Cell Metabolism*, researchers from Princeton University published "The Small Intestine Converts Dietary Fructose Into Glucose and Organic Acids."[8] The authors state:

> while it is commonly believed that the liver is the main site of fructose metabolism, their research shows that it is actually the small intestines that clears most dietary fructose, and this is enhanced by feeding. High fructose doses spill over to the liver and to the colonic microbiota.[9]

However, this study was conducted on mice, and not humans.

Studies "linking" HFCS to various negative health consequences are essentially cherry-picking HFCS as the cause vs. the individual's lifestyle in general. Since HFCS appears mostly in many processed foods, a.k.a. junk foods, anyone consuming high intakes of HFCS is also likely on an LSD diet, which is not the psychedelic drug popular in the 1960s, but the Lousy Stinking Diet I have already mentioned. As I have stated elsewhere, this type of diet is likely one very low in fruits, vegetables, and whole grains. To my freshman students, this is also referred to as an atherogenic diet, or the "I do not want to live too long diet." This diet is one that promotes a myriad of other negative health conditions like obesity, which HFCS is frequently accused of causing. So, it is the overall lifestyle of the individual which is the culprit, and not HFCS. The high intake of HFCS is simply reflective of a poor lifestyle, which of course precipitates the negative health outcomes. This is simply common sense. The molecule of HFCS is readily metabolized like any other sugar molecule, and it is not the presence of HFCS which is the problem, but the overindulgence of foods where most of the HFCS appears. These foods include sodas, candy, baked goods, snack foods, energy drinks, etc. Attempting to pin a cause-and-effect relationship of HFCS intake to obesity, or any negative health condition, is junk science.

Salt

Salt is added to foods for both preservation as well as flavor, and has a wide variety of important biological roles. The current federal recommendations for daily salt intake are highly controversial due to its level of restriction for most consumers, which some have argued may hinder many of salt's necessary biological functions. There are certainly those consumers who, due to genetics and other possible medical conditions, may need to adhere to such restrictive guidelines, but for most consumers, the current restrictive guidelines of no more than 2,300mg/day are not warranted and may lead to unrealistic restrictive food intake. In an August 14, 2014 editorial in the *New England Journal of Medicine*, "Low Sodium Intake—Cardiovascular Health Benefit or Risk?," it is stated that current research results "argue against the reduction of dietary sodium as an isolated public health recommendation."[10]

Further, the 2018 *British Medical Journal Nutrition, Prevention & Health* review includes "Excessive Dietary Sodium Intake and Elevated Blood Pressure: A Review of Current Prevention and Management Strategies and the Emerging Role of Pharmaconutrients," which states:

> while elevated blood pressure is an established surrogate marker for CVD and a risk factor meriting pharmacological and dietary interventions, what tends to be ignored in discussions on this topic is the concept of salt sensitivity and the difference in response individuals can have to sodium intake. Considering the salt sensitivity phenotype is important since many dietary interventions aim to decrease sodium intake to lower blood pressure, but this assumes individuals are all salt sensitive, which is not the case for many.[11]

So, unless you are salt sensitive, an excessively restricted sodium intake in unwarranted.

Again, the problem here is a far too heavy reliance on association studies indicating individuals with the high sodium intakes experience a problematic rise in hypertension. This is true. However, this is far more likely related to the LSD diet already discussed and the related sedentary lifestyles most consumers embrace, instead of a direct association with salt alone. Those with an excessively high salt intake would have to consume a significant amount of processed foods to achieve the level of intake which is problematic. To do so simply illustrates a poor diet in general. If you have a high intake of extremely salty foods, you are going to simultaneously have a low intake of fruits, vegetables, and whole grains. Again, common sense.

The important point here is that if you are salt sensitive, salt restriction would be a useful tool in the prevention of hypertension and cardiovascular disease. However, if you are not salt sensitive, the current dietary guidelines for sodium are likely too restrictive and unnecessary. An excellent review of this topic is provided by the American Council on Science and Health in the publication *Does Excess Dietary Salt Cause Cardiovascular Toxicity?*[12]

Artificial Colors

There are only seven artificial colors approved by the FDA which have been extensively tested. However, the public perception of synthetic food colors has been significantly influenced by anecdotal reports or association studies unsupported by scientific evidence, such as increased hyperactivity in children. This anecdotal evidence also fuels the motivation of some purported consumer groups, such as the Center for Science in the Public Interest (CSPI), to petition the FDA to request such companies as Kraft Foods to discontinue using two artificial colors Yellow No. 6 and 5 in their mac and cheese products.

However, anecdotal evidence is a personal account of something that has occurred due to your exposure to it. As an example, anecdotally, many parents will claim, and many still believe, that sugar makes their kids hyperactive. However, when these same children receive the same product made with an artificial sweetener, which will have no impact on blood sugar levels, the parents still perceive, anecdotally, their children as hyperactive. The real initiator for the hyperactivity of the children is the ingestion of a product they enjoy, which in turn can readily influence behavior as a result. The behavioral change is related to pleasure or the experience of the taste of the product, not the ingredient. A hungry and lethargic child will obviously experience a considerable change in their behavior with a belly full of tasty mac and cheese, which is an unrelated effect of the coloring agent.

Additionally, artificial colors are far cheaper to produce and have a much longer shelf life than their natural counterparts, both benefits to consumers. They also help maintain the color of the product desired by the consumer. The natural color of the product would naturally deteriorate with exposure to sunlight, air, temperature extremes, moisture, and storage conditions, which in turn would deter any consumer from any desire to purchase or consume the product.

Artificial Flavors

In 2015, a coalition of purported consumer and health advocacy groups petitioned the FDA to remove seven synthetic flavoring agents from the marketplace, which took place in October 2018 and will be enforced in 2020. According to the Chemical and Engineering News (C&EN) October 9, 2018 report *FDA Bans 7 Synthetic Food Flavorings,*[13] the petitioners provided the FDA with evidence that the flavorings are carcinogenic in laboratory animals. However, C&EN points out that the "FDA claims that the chemicals do not pose a risk to public health under the conditions of their intended use, but the agency cannot legally authorize the use of food additives that have been shown to cause cancer in animals."[14] Most anything will cause cancer in test animals if given in high enough dosages, dosages so high that it would be impossible to achieve with normal human consumption.

However, here is the ultimate irrationality of this ban which is reflective of the whole chemophobia issue. Each of the synthetic flavoring agents has a natural counterpart that can be used and is not affected by the ban. In other words, if you can extract the same exact chemical from a natural source as you can by producing it synthetically, you can use the flavoring agent in food, even though it is the same exact chemical. The purported risk of the chemical has not changed; just the source of it. This will directly affect the cost of the product to the consumer due to the higher cost of extracting the same chemical from food vs. making the synthetic version.

Dr. Josh Bloom, the Director of Chemical and Pharmaceutical Science for the ACSH, provides the following examples illustrating how irrational this decision is using three of the seven chemicals banned. The following examples are taken from Dr. Bloom's May 29, 2018 article *Enviro-Thugs Sue To Keep Natural Food Flavors Out of Food.*[15]

- **Myrcene:** Occurs naturally in lemongrass, hops, basil, and mangos, and ginger.
- **Pulegone:** Is another naturally occurring chemical that is found all over the place. The minty-smelling chemical is found in peppermint, spearmint, blueberries, chamomile, black currant, catnip, black tea. . . . Like myrcene, pulegone is listed as a carcinogen on Crazifornia's Prop 65 list.
- **Methyleugenol:** Is a clove-scented chemical that is also found widely in plants and fruits, for example, nutmeg, apples, banana, orange juice, grapefruit, among others.

MSG

Monosodium glutamate (MSG) has had a long-standing undeserved poor reputation as a flavor enhancer. The compound itself is simply a combination of sodium and the naturally occurring amino acid L-glutamate, which is found in any food with protein, as well as naturally produced by our bodies. MSG is just the salt form of this amino acid, and a small fraction of the population may be sensitive to it. This may make the chemical problematic to a select few due to genetics, but it is irrelevant to most. After the consumption of MSG, the molecule is split into sodium and glutamate, both having well-established normal metabolic pathways. The glutamate segment is just a fraction of what one would normally consume daily in a typical diet and glutamate is a fuel source for the cells which line the digestive tract. The sodium segment of the molecule would be metabolized just like any other source of sodium from the diet. For additional safety information for MSG, I suggest the FDA's "Questions and Answers on Monosodium Glutamate (MSG)," which can be found here: www.fda.gov/Food/Ingredients PackagingLabeling/FoodAdditivesIngredients/ucm328728.htm.

Brief videos of the safety of MSG can be found here:

- American Chemical Society: "Is MSG Bad for You? Debunking a Long-Running Food Myth": http://youtu.be/VJw8r_YWJ9k.
- American Council on Science and Health: www.youtube.com/watch?v=uW0_A74bENM.

Purported Endocrine Disruptors

There are frequent claims of various chemicals in the food supply which are often erroneously referred to as endocrine disruptors, which has no real scientific meaning outside of food activists' attempts to instill fear in consumers. Bisphenol A (BPA), which is a chemical used in many food container linings, is one of the most targeted by food activists. BPA forms a barrier between the food itself and the can. The separation of the two prevents the metal itself from corroding and the transfer of the metal to the food. Both properties help extend the shelf life of the product significantly (sustainable agriculture). Your exposure to BPA is roughly 1,000 times below the established safe limits. Based upon data from the CDC, the typical exposure level to BPA for the average consumer is only 2.4 micrograms per day. The *Factsaboutbpa.org,* which is part of the American Chemistry Council, put this exposure level into perspective in their publication *What Does Government Research Tell Us About BPA?*[16] "If a small mint was broken into pieces that each weighed 2.4 micrograms, it would take 240 years to consume all of the 87,500 pieces at a rate of one piece per day." Also, "a person would have to ingest 1,300 pounds of food and beverages in contact with polycarbonate plastic every day to exceed the safe limit." Additionally, BPA is rapidly converted in the intestines to a substance with no biological activity and eliminated from the body in the urine within 24 hours. BPA purported ability to bind to estrogen receptors is 12,500 times below that of estradiol, and there is no evidence that this weak activity has any effect on the endocrine system.

For a brief FDA safety report of BPA, go to: www.fda.gov/Food/IngredientsPackagingLabeling/FoodAdditivesIngredients/ucm355155.htm, or these videos from the ACSH:

- "The Science of the 'Endocrine Disrupter' Debate" explains the absence of science: https://youtu.be/8jm9SE-AY7I.
- "BPA: The False Poster-Child of Bad Chemicals": www.youtube.com/watch?v=8BGSgOftnmw.

From the CropLife.org review "Endocrine Science":[17]

> With ongoing public discussion of endocrine disruptors (EDs), EndocrineScienceMatters.org was launched by CropLife International to publicly discuss the latest science on key issues and how crop protection products are tested for potential endocrine-disrupting effects.
>
> Strong scientific evidence shows that crop protection products do not cause endocrine-related diseases or conditions such as cancer, diabetes or obesity. In fact, human exposure to these products is orders of magnitude lower than exposure to common, natural and more potent endocrine-active substances like sugar, caffeine and soy protein. Independent of such

substances, multiple factors account for increases in endocrine-related diseases or conditions, such as lifestyle, diet, body weight and changes in diagnostic criteria.

From *Toxicology Letters*, "Endocrine Disruption: Fact or Urban Legend?":[18]

Overall, despite of 20 years of research a human health risk from exposure to low concentrations of exogenous chemical substances with weak hormone-like activities remains an unproven and unlikely hypothesis.

From *Policy Focus*, "The Science of the 'Endocrine Disrupter' Debate," January 2014:[19]

Americans are increasingly being told by the media and environmental activists that common consumer goods—from plastics to cosmetics to flame retardant-furniture—contain chemicals that endanger their health. These chemicals are referred to as "endocrine disrupters." The activists charge that they affect our hormones, cause cancer, harm our children's health, affect fetal development, and even make us fat. Many of these campaigns are targeted at women, particularly mothers, who naturally are concerned.

However, there is no evidence to suggest that the chemicals in these consumer products actually have such effects on humans at current exposure levels. In fact, these chemicals are far too weak and human exposure too low to produce any measurable impacts. Moreover, similar, naturally occurring chemicals are found in many foods and are far more potent than synthetic chemicals, and yet humans safely consume them every day. Accordingly, there is little reason to fear such trace chemicals in consumer products.

These alarmist headlines, however, do result in harm to consumers: they lead to unnecessary regulations and decisions by manufacturers that lead to higher prices, fewer choices, and inferior products. Indeed, products may ironically become less safe as a result of this dynamic as manufacturers substitute away from known, effective chemical additives and use less tested, less effective alternatives.

Consumers ought to get the facts, ignore the alarmist headlines, and discourage regulators and producers from taking action based on groundless fears.

In the section under Naturally-Occurring 'Endocrine Disrupters' in Food, it states:

The entire theory that manmade chemicals are causing significant endocrine disruption falls apart when you consider exposures to naturally occurring endocrine-mimicking chemicals. Plants naturally produce such chemicals called phytoestrogens, to which we are exposed at levels that are thousands of times higher than those of synthetic chemicals. Human exposure levels to synthetic estrogens is minute, particularly when compared to that of naturally occurring estrogens found in fruits and vegetables.

As researcher Jonathan Tolman points out, humans consume these naturally occurring endocrine-mimicking chemicals every day without ill effect. In fact, he explains that tests have found such chemicals in 43 foods in the human diet such as soy, which is used in hundreds of products that we safely consume on a regular basis. Phytoestrogens like those found in soy and other foods are 1,000 to 10,000 times more potent than synthetic estrogens, and the estrogenic effects of the total amount we consume are as much as 40 million times greater than those of the synthetic chemicals in our diets. The data strongly indicate that there is little reason to worry about the impact of trace chemicals on human endocrine systems. Unfortunately, misinformation will continue to guide this debate and public policy until consumers access and respond to more balanced information.

Finally, from ChemicalSafetyFacts.org, "How Do Chemicals Impact the Endocrine System: It Depends on Dose & Duration of Exposure"[20]

Many substances and activities can impact the endocrine system—stress, food, and exercise, for example, are among the most common. What scientists typically focus upon with regard to so-called endocrine-active substances is whether exposure can lead to an adverse health effect.

We know that at high doses and durations of exposure, some substances can negatively impact the endocrine system and cause adverse health effects. At lower doses or limited exposures, however, those substances may not impact our health at all.

How a chemical interacts with the endocrine system depends on a variety of factors, including:

- *Type and duration* of the exposure to the chemical,
- *Frequency* of exposure,
- *Potency* of the substance, and
- How the body *absorbs and eliminates* a substance.[17]

Scientists have extensive knowledge, from decades of research on both natural and synthetic chemicals, about how various exposure levels can cause different effects. It is well established in scientific communities that while a particular substance can benefit people at the right dose, the same substance could cause a different response, including harm, at higher doses. A simple example is aspirin. Taking two aspirin will likely eliminate your headache; taking two bottles of aspirin could eliminate your life.

This principle of dose and response applies to all chemicals, natural and synthetic, used in a variety of applications—from cosmetics and personal care products, to pharmaceuticals, to crop protection and industrial manufacturing chemicals, among others.

Sodium Benzoate

One of the major points of this chapter was to illustrate how false phobias against "hazardous" chemicals are created to enhance the marketing campaigns of various purported "natural" or "clean" food products. This last example receives a five-star rating for deception and was produced by Panera Bread in 2017. Panera Bread produced a "Preservative with Purpose" campaign highlighted in this video: www.panerabread.com/en-us/our-beliefs/preservatives-with-purpose.html.

Panera Bread's advertisement involved sodium benzoate, which is used as a preservative in many foods and is simply the sodium version of the naturally occurring chemical benzoic acid. Panera Bread states in the caption below the video:

> Sodium Benzoate is an artificial preservative found in sauces, jellies and pickled foods. It's also an active ingredient in fireworks. So, to celebrate our removal of all artificial preservatives from Panera food, we staged an Independence Day firework show in Johnston City, Illinois. The town hadn't seen fireworks in 10 years.

Of course, Panera Bread wishes those watching the advertisement will assume that Panera Bread is all about "natural" food ingredients with no "nasty" preservatives, which purportedly makes their food better for you. This is not true. In food, sodium benzoate is used to inhibit the growth of mold, yeast, and bacteria. In other words, it delays spoilage and extends the shelf life and safety of products where it is used. Again, this is part of sustainable agriculture by reducing food waste. Natural food sources of benzoic acid include fruits, vegetables, cheeses, and yogurt, as well as other foods. Sodium benzoate is perfectly safe and rapidly excreted from the body.

Derek Lowe, Ph.D. in organic chemistry, stated the following in his article "Sodium Benzoate Nonsense" in his response to Panera Bread's self-serving advertisement on July 24, 2017:[21]

> Panera's ad is a cute graphic about how sodium benzoate is found in fireworks, so it shouldn't be in your delicious food. The problem is, a goodly number of Panera's menu items—such as all the ones with cheese, and all the ones with berries—contain plenty of sodium benzoate already, in some proportion with benzoic acid. It's stupid and disingenuous of them to pretend that they're protecting their customers from evil industrial chemicals, when the same stuff is found in their own ingredients. As many readers will

appreciate, you can play the same game with all sorts of other ingredients. Lactic acid (found in milk) is used in tanning leather. Palmitic acid, found in meats, coconut oil, sunflower seeds and many other foods, is used in making soap. 2,3-butanediol (a flavor component of many cheeses) is used in making printing ink and as antifreeze. I could go on all day; any organic chemist could. The entertainment value goes down after a while, because the fundamental premise (Good Healthy Natural Stuff versus Toxic Sludge) is stupid to start with.

So, Panera, you're playing on people's lack of knowledge of chemistry in order to make yourselves look good. Your reasoning is faulty, and your science is wrong. Your ads are offensive to anyone who actually understands chemistry, not that you care much, and you're claiming a halo for yourselves that you don't have.[22]

The title of this chapter is "Chemophobia and The Boy Who Cried Wolf." In the fairytale "The Boy Who Cried Wolf," it only took two times for the young shepherd boy to mislead the villagers before the villagers would no longer respond to his false cries. One can only assume that the villagers in this tale were not as gullible as many consumers who repeatedly allow themselves to be victimized by baseless fear-mongering of the purported health threats of safe chemicals used to provide them their food, regardless how many times these claims have turned out to be false.

It doesn't matter how beautiful your theory is, it doesn't matter how smart you are. If it doesn't agree with experiment, it's wrong.[23]

Richard P. Feynman (Nobel Laureate Physics, 1965)

The take-home message from this chapter is as follows: consumers should not fear chemicals in the food supply. They provide numerous benefits, as stated, and without them, our food choices would be greatly restricted. It is not the chemicals consumers are consuming which are the problem; it is the chemicals consumers are not consuming. Most diets, in any developed country, are woefully low in produce and whole grains, resulting in very low intake levels of thousands of plant chemicals (phytochemicals), most of which likely play a significant role in our health. Most studies which purport to link a chemical compound to any myriad of diseases or physical ailments, has, in most cases, simply cherry-picked a chemical compound which may be in high use among a population group who is also adhering to overall poor lifestyle and dietary habits. It is not the chemical which is the problem; it's the lifestyle and chronic poor food choices—so find another phobia. Chemicals in the food supply are not the problem; lifestyle choices are more likely the problem.

For a more thorough review of the chemophobia issue, I recommend the upcoming updated documentary produced by the American Council on Science and Health, *Big Fears Little Risk*.[24] ACSH promotional material for the documentary states:

> "Big Fears Little Risks" is a documentary, but unlike most of what you see on places like Netflix, it is pro-science, and not scaremongering trace chemicals, food, or the modern world.
>
> Instead, we note that humans are generally bad at evaluating risk while handing over much of our decision-making to "bubbles of eminence" in self-appointed groups who create guidelines; if the American Academy of Pediatrics says we are all bad parents by letting our children walk to school alone, many of us accept that. If organic food trade groups or their advocates say buying organic food will get your child into Harvard, some will do so—because the only cost is money.
>
> But for many poor people, money is not a second order issue—a war on science and technology is a war on the poor, because they are impacted most by higher costs that derive from meaningless distinctions.
>
> "Big Fears Little Risks" makes it safe to eat breakfast again, no matter how many trial lawyers in California want to sue over trace ingredients that have only ever harmed rats, and only at high doses.

Following is one example of how the media's misleading coverage of science news drives the chemophobia issue.

Media Example 1

Headline:

> *Newsweek:* "Your Fruit Is Covered with Nasty Pesticides: Scientists Have Discovered the Best Way to Wash Them Off" (October 25, 2017).

The *Newsweek* article was based upon a report published in the *Journal of Agricultural and Food Chemistry*,[24] which had provided the results of washing pesticide residue off apples using three different methods. These methods were Clorox bleach, baking soda, and tap water. The researchers, from the University of Massachusetts, stated that "removal of pesticide residues from fresh produce is important to reduce pesticide residue exposure to humans,"[25] and concluded that washing apples with baking soda was the most effective of the three methods. However, here is the rub.

The researchers applied the maximum legal limit of pesticides onto the apples, which would be expected, but only waited 24 hours prior to washing and measuring for residue samples after attempting the three methods of cleaning. In real life, the EPA does not allow any harvesting of apples after spraying for seven days, not 24 hours, and 14 days if it is a "pick yourself orchard." This allows considerable time for any pesticide applied to degrade by normal environmental conditions and reach well-established safe levels of exposure (see Chapter 4). As an example, thiabendazole, the fungicide used in this study which is used to prevent mold growth during the storage of apples, has a legal tolerance set at 5 parts per million (ppm). Here are some examples from the American Council on Science and Health as to what 1ppm looks like in real world terms:

- 1g of residue in 1,000,000g (2,200 pounds) of food.
- 1 inch in 16 miles.
- 1 minute in 2 years.
- 1 cent in $10,000.
- 1 pancake in a stack 4 miles high.

So, the legal 5ppm is equivalent to 5 grams of residue in 2,200 pounds of apples. However, even this insignificant level is not what is exposed to consumers. Steve Savage, who holds a Ph.D. in plant physiology from UC Davis, operates the website Applied Mythology. He pointed out in an email dated October 28, 2017 to me, that "the 2014 USDA data shows that the residues found are all below the 5-ppm tolerance." He forwarded this data, and this is what I found: 88% of the samples detected are far less than 1ppm of residue, or less than 1 gram of residue for every 2,200 pounds of apples. Twelve percent of samples detected only 1–2 ppm, or 1–2 grams for every 2,200 pounds of apples. This is an incredibly insignificant amount. But just how insignificant is this, and what does it mean in real life terms?

Thiabendazole is referred to as a systemic pesticide, which means it is absorbed into the tissue of the produce where it is applied. The absorption is the intended result of the pesticide. This is how it prevents the fungal growth during storage and allows consumers who do not have an orchard in their backyard, or close by, to have access to and enjoy apples. However, it is rapidly excreted in the urine and feces. It does not bioaccumulate in body tissue. As Dr. Savage points out, "without these fungicides applied on the way into storage, there would be much more food waste (non-sustainable food production). Therefore, one would expect to find residues." So, the obvious point here is the standard risk-to-benefit ratio with the use of thiabendazole. I would argue that it is quite clear that considering the non-existent health risk for its use compared to the health benefits (the storage and shipping of apples to millions of people), we need to be thankful for the product's development.

The other pesticide used on the apples was phosmet, which is used to help control for apple maggot, something no one wants in their apple when they bite into it. Phosmet does not penetrate the produce flesh, and the EPA mandates a waiting period of seven days post-application before apples can be harvested, not 24 hours as used in this study. The EPA has set a tolerance level of 10ppm, which is 10 grams per 2,200 pounds of apples. However, as with all pesticides, this is not what the consumer is exposed to, if any. The 2014 USDA sampling data demonstrate detection levels of 0.005–0.280ppm. Again, these are incredibly low numbers, almost a non-existent detection level. Most samplings were below the 0.076ppm levels, and this would be prior to the consumer rinsing them off with water at home.

Now, considering the facts and a little more detailed evaluation of actual exposure to the pesticides (chemicals) used on apples, do you consider the *Newsweek* headline "Your Fruit Is Covered with Nasty Pesticides" to be fake news or real news? Enjoy your apple pie and be thankful for the lack of mold and maggots in the apples you used to make it with.

Notes

1. *Food and Agriculture Organization of the United Nations. Report: The State of Food Security and Nutrition in the World*, 2019, p. 2. http://www.fao.org/state-of-food-security-nutrition/en/
2. *CDC Facts About Cyanide*, p. 1. https://emergency.cdc.gov/agent/cyanide/basics/facts.asp
3. *American Council on Science and Health Holiday Dinner Menu*, pp. 4–5. https://www.acsh.org/sites/default/files/Holiday%20Menu%20final%202004.pdf
4. *The Naturalness Fallacy*, by James Kennedy, M.S., p. 17.
5. *NIH Office of Dietary Supplements, Vitamin B6 Fact Sheet for Health Professionals*. https://ods.od.nih.gov/factsheets/VitaminB6-HealthProfessional/
6. *Assessing Toxic Risk (Cornell University Scientific Inquiry Series—Student Edition Publication)*, by Nancy M. Trautmann, p. 7.
7. S.Z. Sun and M.W. Empie, *Fructose metabolism in humans – What isotopic tracer studies tell us. Nutrition and Metabolism (Lond)* (October 2, 2012), Vol. 9, No. 1, p. 89.
8. C. Jang, S. Hui, W. Lu, et al., The small intestine converts dietary fructose into glucose and organic acids. *Cell Metabolism* (2018), Vol. 27, No. 2, pp. 351–61.
9. Ibid, p. 2.
10. S. Oparil, Low sodium intake—Cardiovascular health benefit or risk? *New England Journal of Medicine* (2014), Vol. 371, No. 7, pp. 677–79. http://dx.doi.org/10.1056/NEJMe1407695
11. O.M. Dong, Excessive dietary sodium intake and elevated blood pressure: A review of current prevention and management strategies and the emerging role of pharmaconutrigenetics. *BMJ Nutrition, Prevention and Health* (2018), p. 5.
12. https://www.acsh.org/sites/default/files/Does-Excess-Dietary-Salt-Cause-Cardiovascular-Toxicity.pdf
13. Brett E. Erickson, FDA bans 7 synthetic food flavorings. Agency responds to evidence that chemicals can cause cancer in animals. *Chemical & Engineering News* (October 9, 2018), p. 1.
14. Ibid, p. 1.
15. https://www.acsh.org/news/2018/05/29/enviro-thugs-sue-keep-natural-food-flavors-out-food-13020
16. www.factsaboutbpa.org/what-does-us-government-research-tell-us-about-bpa
17. https://croplife.org/crop-protection/endocrine-science-matters/
18. G.J. Nohynek, C.J. Borgert, D. Dietrich, and K.K. Rozman, Endocrine disruption: Fact or urban legend? *Toxicology Letters* (December 16, 2013), Vol. 223, pp. 295–305. https://www.ncbi.nlm.nih.gov/pubmed/24177261

19. https://c1355372.ssl.cf0.rackcdn.com/94953b38-031b-473a-b5fd-30280c172a8f/PolicyFocus15_Jan_p2.pdf
20. https://www.chemicalsafetyfacts.org/chemistry-context/chemical-exposures-and-our-endocrine-system/how-do-chemicals-impact-the-endocrine-system/
21. https://blogs.sciencemag.org/pipeline/archives/2017/07/24/sodium-benzoate-nonsense
22. Ibid, p. 4.
23. Richard P. Feynman (Nobel Laureate Physics, 1965) 1964 lecture at Cornell University – The Key to Science. https://www.youtube.com/watch?v=EYPapE-3FRw.
24. https://www.acsh.org/big-fears-little-risks-help-fund-our-documentary
25. T. Yang, J. Doherty, B. Zhao, A.J. Kinchla, J.M. Clark, and L. He, Effectiveness of commercial and homemade washing agents in removing pesticide residues on and in apples. *Journal of Agriculture and Food Chemistry* (October 25, 2017), Vol. 65, No.44, pp. 9744-52. https://pubs.acs.org/doi/10.1021/acs.jafc.7b03118

4

THE FABRICATED ORGANIC FOOD MARKET

For the last 15 years, I have taught an introductory nutrition course at a local junior college. On the opening night, one of the many questions I ask students as part of the introductory information to the course is how many of them or their immediate family members are directly involved with agriculture. The response is always the same. Either no one responds, or, if they do, it is always less than 2% of the class. This is in a community, Kern County of California, which was ranked as the number one agricultural producing county in the United States in 2017, at $7.1 billion,[1] as I mentioned in an earlier chapter.

Nationally, according to the US Environmental Protection Agency Ag 101, which is a brief overview of American agriculture, only 1%–2% of the population is directly involved in the growing and producing the food for what the other 98% need. The other 98% are free to choose other livelihoods. This results in most consumers having limited to no understanding regarding standard agricultural practices, basic plant physiology, and the global food challenge to feed over 7 billion people. These consumers often embrace an irrational fear of chemicals, as illustrated in Chapter 3. This unfounded consumer fear remains despite the fact that all things, including themselves, are made up entirely of chemicals. This leaves 98% of the population as easy targets to victimize with a wide variety of misinformation intended to exploit their misunderstandings. This section, as well as Chapter 3, illustrates this.

As an example, consider this simple illustration. Using the USDA Organic label as shown in Image 4.1, I ask students a simple question: based upon your perception of what this label is supposed to indicate to consumers, are the produce items labeled as such inherently safer and more nutritious than conventionally grown? The overwhelming response is always yes.

IMAGE 4.1 USDA Organic seal.

However, the USDA specifically states that the seal has nothing to do with "food safety or nutrition," which are the main selling points of organic crops to consumers. The label simply indicates that the produce was purportedly grown adhering to certain rules and regulations and has nothing to do with any safety or nutritional benefits. In 2000, when this label began being distributed by the USDA, Dan Glickman, the Secretary of US Department of Agriculture at the time stated, *"let me make it clear about one thing. The organic label is a marketing tool. It is not a statement about food safety. Nor is 'organic' a value judgement about nutrition or quality"* [my emphasis].[2]

On April 8, 2014, Professor Bruce Chassy, Ph.D., professor emeritus at the University of Illinois, Department of Food Science & Human Nutrition, published "Why Consumers Pay More for Organic Foods? Fear Sells and Marketers Know It."[3] The article appeared in *Academic Reviews*, an independent 501(c)(3) nonprofit organization of academic professors and researchers from around the world who establish sound science in food and agriculture. He states:

> Our report finds consumers have spent hundreds of billion dollars purchasing premium-priced organic food products based on false or misleading

perceptions about comparative product food safety, nutrition and health attributes. The research found extensive evidence that widespread, collaborative and pervasive industry marketing activities are a primary cause for these misperceptions. This suggests a widespread organic and natural products industry pattern of research-informed and intentionally-deceptive marketing and paid advocacy. Further, this deceptive marketing is enabled and conducted with the implied use and approval of the U.S. government endorsed and managed U.S. Department of Agriculture (USDA) Organic Seal and corresponding National Organic Standards Program (NOSP) in direct conflict with the USDA's NOSP stated intent and purpose.[4]

John Block, Secretary of the US Department of Agriculture from 1981–1985, wrote a commentary on the findings of the *Academic Reviews*. In this commentary for the *Genetic Literacy Project—Science Not Ideology* on May 16, 2014, he states, "the organic seal does not and cannot signify any health or safety criteria whatsoever. It merely certifies that products were produced using less modern inputs."[5] He goes on to state:

> I am supportive of those who choose to grow organic food and those who choose to buy it. However, I do not accept the organic industry's attack on new tech agriculture. It is entirely without justification. If the roles were reversed and conventional agriculture engaged in similar "black marketing" against organics, the regulatory authorities and consumer groups would come down like a ton of bricks.[6]

He closes his comments with, "one wonders why, then, this multibillion-dollar industry gets a free ride to propagate negative and false advertising denigrating the livelihood of the vast majority of America's farmers. It's time for it to stop."[7]

In short, the organic industry knows that unless it can effectively demonize conventionally grown produce by creating a consumer misconception of superior safety and health of organics by using blatantly false and deceptive marketing methods, then the foundation of the whole organic food industry and its ability to enhance sales would be impossible. Consumers would simply be unable to justify the additional cost of the produce or products based upon the facts.

Prior to addressing the five misunderstandings related to organic food industry, consumers should recognize the international problem associated with feeding the world's population.

1. As of 2019, there are roughly 7.7 billion mouths to feed.
2. Human food production competes with 600 species of insects, 1,800 species of plants (weeds), and numerous species of fungi and nematodes.[8]
3. According to the USDA/National Institute of Food and Agriculture report "Global Scientists Meet for Integrated Pest Management Idea Sharing,"

published April 14, 2015, "In the developing world, 40–50 percent of all crop yields are lost to pests, crop diseases, or post-harvest losses. Even in the United States, that number is 20–25 percent."[9]

4. There is limited farmland for production, so yield per acre will become an increasingly significant concern.

Now, there are five misunderstandings regarding organic foods which many consumers embrace. These misunderstandings lead them to believe that they must spend roughly 40% more for their food.

These misunderstandings are:

1. Organic foods are healthier, which they are not.
2. Organic foods are safer, which they are not.
3. Organic foods taste better, which they do not if both products are picked equally ripe.
4. Organics are inherently better for the environment, which is not true.
5. Is it even organic in the first place?

Misunderstanding 1: Organic Foods Are Healthier

Millions of consumers believe that organics are healthier than conventionally grown foods. But is this belief true? No—it defies basic plant physiology. Plants are autotrophic, which means they can utilize the basic elements of the soil, such as minerals and water, as well as carbon dioxide from the air and sunlight, and, through photosynthesis, synthesize the wide gamut of molecules they need to support their growth. This includes all vitamins. The vitamin content of a crop is going to be based upon the plant's genetics and the conditions it is grown in. If the growing conditions are the same and soil composition equivalent, a tomato's nutrient content, whether grown "organically" or conventionally, will be the same.

The tomato plant's root system will just as readily absorb and utilize synthetic minerals from fertilizers as it would minerals from manure or other forms of compost. An easily understood analogy would be a female diagnosed with anemia due to iron deficiency. Does her treating physician tell her to go home and consume a large amount of red meat, an organic source of iron, or simply prescribe a synthetic source of the iron in the form of a supplement (fertilizer)? Her small intestines will not differentiate between the two sources of iron, so it truly does not matter. The absorptive surface area of her small intestines, as the absorptive surface area of the tomato's roots, will readily absorb either source because the structure of the molecule it needs—in this case, iron—is the same in both cases. The tomato root system is looking for a specified structure, not the source of it. This basic plant physiology is why no well-controlled study utilizing identical growing conditions will ever find any significant nutrient differences between organics and conventionally grown crops.

On February 4, 2000, I watched a broadcast on ABC's *20/20* presented by John Stossel regarding the organic food industry and its claims of superiority to conventionally grown foods. Mr. Stossel interviewed the top official of the American organic industry, Katherine DiMatteo, Executive Director of the Organic Trade Association (OTA) at the time. When Ms. DiMatteo was asked if organic food is more nutritious than regular food, she responded: "It's as nutritious as any other product" (twice). Essentially, she is simply saying no. When asked if organics were safer than regular food, she again responded truthfully. "Organic agriculture is not particularly a food safety claim." Again, she is essentially saying no.

The same broadcast was reviewed by Dennis Avery in *Priorities for Health* in 2000, "Besieging Stossel to Protect Organic Myths."[10] *Priorities for Health* is a publication of the American Council on Science and Health (ACSH), one of the most well-respected consumer education sites in the United States. Mr. Avery is Director of Global Food Issues for The Hudson Institute of Indianapolis. He stated that Ms. DiMatteo:

> was forced to admit on national television that there is no evidence for the industry's long-standing contention that organic foods are safer and nutritionally better than their mainstream counterparts. In effect, she conceded that the industry's whole campaign against mainstream foods has been a lie.[11]

Here are a few examples illustrating no nutrient related differences between organics and conventional crops.

In 2014, the Penn State Food Safety blog "Organic Food Safety—Fact Versus Hype," written by John Block, the US Secretary of Agriculture from 1981 to 1985 stated, "Is organic food more nutritious than conventionally grown food? No. There is no evidence to support this."[12]

In 2012, the *Annals of Internal Medicine* reported on the results of a Stanford study, "Are Organic Foods Safer or Healthier Than Conventional Alternatives? A Systematic Review." The Stanford researchers concluded that the research does not support such a claim.[13]

In 2010, The *American Journal of Clinical Nutrition* published "Nutrition-Related Health Effects of Organic Foods: A Systematic Review." The authors identified 12 relevant studies and concluded, "from a systematic review of the currently available published literature, evidence is lacking for nutrition-related health effects that result from the consumption of organically produced foodstuffs."[14]

In 2009, the *American Journal of Clinical Nutrition* published "Nutritional Quality of Organic Foods: A Systemic Review." The researchers reviewed the published research available, which included 162 studies from January 1, 1958 to February 29, 2008, 55 of which were of satisfactory quality and used for this review. Their conclusion: "there is no evidence of a difference in nutrient quality between organically and conventionally produced foodstuffs." However, the remaining 107 studies, which were poorly controlled, are reflective of irrelevant information due to too many uncontrolled variables or poor study design. These studies are often the very ones cherry-picked by organic enthusiasts attempting to enhance their image.

In 2008, the Danish Research Centre for Organic Farming funded a study to determine the nutritional value of organics vs. conventional foods by the Department of Human Nutrition at the University of Copenhagen. The results were published in the Society of Chemical Industry's (SCI) *Journal of the Science of Food and Agriculture*.[15] Here are the main points of that study:

- Five different crops were studied—carrots, kale, mature peas, apples, and potatoes.
- They were cultivated both organically and conventionally.
- Study leader Dr. Susanne Bugel said: "No systemic differences between the crops," so the study does not support the belief that organically grown produce is nutritionally superior.[16]

Notice the sponsor of this study, the Denmark organic industry itself, illustrating that conventional food farmers did not head this study to attack organic farmers, but was in fact funded by the organic industry.

In 2007, the British Nutrition Foundation published its conclusions which stated that overall the body of evidence does not support the view that organic food is more nutritious than conventionally grown.[17]

Findings presented at a 1997 Tufts University conference also illustrated that organic food did not prove significantly superior nutritionally in any of the comparative tests of organic food and regular food conducted in 19 countries.[18]

However, let's say theoretically that an organically grown orange was shown to contain 10% more vitamin C than a conventionally grown one. A large orange contains roughly 97mg of vitamin C, which is more than enough to maintain the maximum storage capacity of vitamin C of 1,500mg. Now, theoretically, your organic orange would contain 10% more than this, which would be roughly 107mg. In Chapter 8, I illustrate that vitamin C intake above 100mg would begin to have a significant excretion rate due to the body's inability to utilize it or store it. So, the obvious question is this: even if, theoretically, the organic orange contained more vitamin C than the conventionally grown one, would it make any impact on your health? The obvious answer is no. It would be analogous to having a full fuel tank and then continuing to attempt to add more.

Finally, consider the following from "Are Organic Foods Healthier Than Conventional Foods?" provided by the *Genetic Literacy Project—Science Not Ideology*:[19]

"Any consumers who buy organic food because they believe that it contains more healthful nutrients than conventional food are wasting their money," said Joseph Rosen, emeritus professor of food toxicology at Rutgers University.

In the *At a Glance* section:

> The growth in popularity of organic foods has been driven, to a large extent, by claims that they are healthier or more nutritious than those grown by conventional farming methods. Organic boosters argue that the synthetic pesticides and herbicides used by conventional farmers degrade the quality of the soil and result in more pesticide residue at potentially dangerous levels in our food. This is misleading as all pesticides in use today are tightly regulated by the Environmental Protection Agency and pose no threat to human health when used as directed. It should be noted that organic farmers also use pesticides, and often the same chemicals applied by conventional farmers.
>
> Most independent studies indicate that there are no significant health or nutritional differences between food grown conventionally versus organically. There are limited examples of organic crops or conventional crops with greater levels of a particular nutrient, but at levels that are not materially significant. When the relative costs are taken into account, the cost per unit of nutrient, conventional crops come out far ahead. Most nutritionists argue that it is more important for people to increase their consumption of fruits and vegetables—regardless of how they are grown.

In the *The Takeaway* section:

> There is no independently produced evidence in the scientific literature that organic foods offer any consistent nutritional or safety advantages over conventional foods. Some claimed advantages, like higher levels of antioxidants or omega-3 fatty acids or phenols, may not be advantages at all, say scientists. Other differences may be the result of whether a cow was grass fed or grain fed and have nothing to do with whether it was raised organically. Claims that organic crops are more "nutrient dense" have not been consistently supported in independent studies.
>
> On the subject of pesticide residue in foods, scientists agree that conventional foods are more likely to contain higher levels of residues in some cases, but the elevated levels are insignificant. Stanford's analysis, for example, found that organic food had 30 percent lower residues than conventional foods, but all conventional foods had pesticide levels well within global safety standards.
>
> In sum, most crops, including organic foods, contain trace amounts of harmless pesticides. Approved crop protection chemicals used by conventional and organic farmers are safe; the differences in toxicity to humans are insignificant.

Chemical pesticides, whether synthetic or natural, are "essentially non-toxic" when used appropriately. They pose no serious harm to the farmers who use them or consumers who encounter trace residues in their food. As a result, the scientific consensus remains that organic fruits and vegetables are no healthier or safer than their conventional counterparts.

Misunderstanding 2: Organic Foods Are Safer

Although the organic industry comes up with considerable persuasive rhetoric concerning the purported safety of organics, I do not believe you can convince the thousands of people affected by the E. coli-infected, organically grown bean sprouts in Germany in 2011 that organically grown necessarily means safer. According to a March 2, 2012 *Food Safety News* report, "An Outbreak Like Germany's Could Happen Here,"[20] nearly 4,000 cases, 54 deaths, and 845 cases of hemolytic uremic syndrome resulted from the ingestion of *E. coli*-contaminated bean sprouts from an organic farm in Germany. In this report it also quotes Germany's Health Minister as saying, "Of course we know that organic is not really safer."[21]

In 2015, the *Journal of Food Protection* reported on the findings of pathogenic bacteria from 242 samples of various vegetables sold at six different northern California farmers markets. Their findings: "there was a twofold higher probability of Salmonella contamination in samples from growers or vendors who stated they used organic farming practices compared with samples from those using conventional farming practices."[22] This is likely due to the use of manure as fertilizers that was not properly composted for at least a year prior to use. Improperly composted manure increases the risk of bacterial contamination.

The 2014 Penn State blog also stated:[23]

> While there is little support to indicate organic is safer, in some cases, there actually may be increased risk. Organic farmers and processors do not have the arsenal of preventive measures available that conventional farmers and processors do, so spoilage and pests can be a bigger issue. One issue in particular . . . mold spoilage . . . has the potential to increase the risk of mycotoxins, byproducts of mold growth that can cause serious health consequences. Cleaning and sanitizing also becomes more difficult since there are limited choices of what can be used. The same goes with preservatives.

On October 10, 2014, John Block, previously identified, published the commentary "Consumers Are Misled About Organic Safety" for the *Des Moines Register*, again prompted by the findings of the *Academic Reviews* study mentioned earlier. I am providing his full commentary due to his past position as Secretary of the US Department of Agriculture, as well as the significance of his comments for consumers.[24]

> Every day millions of shoppers are paying out as much as 50 or 100 percent more to buy organic foods for themselves and their families. I have friends who make these choices because they have no reason to question claims on labels, in advertising and on social media that organic foods are safer, healthier and more nutritious.
>
> One thing they will not read on any label is a new finding from Academics Review, a group of scientists dedicated to testing popular claims against peer-reviewed science.
>
> The scientists' conclusion based on U.S. Food and Drug Administration (FDA) reported recall information: Organic foods are four to eight times more likely to be recalled than conventional foods for safety issues like bacterial contamination. Nor will consumers see anywhere a reference to the body of peer-reviewed research finding that organic foods are no more nutritious than foods produced by conventional agriculture.
>
> Why are consumers so misinformed? This is not an unimportant problem. It's dangerous. The very people most likely to seek out organic food for its purported safety—the elderly, pregnant women, parents of young children and people with compromised immune systems—are most at risk from organic's higher risk of contaminants, including deadly e-coli.
>
> As Academics Review founder Bruce Chassy, a professor of food microbiology at the University of Illinois, recently reported to a professional trade association, not only is the federal government failing to require that the organic food industry state these risks to consumers. It also allows organic companies to make unfounded safety claims that, if they were made by any other industry, would attract the ire of federal regulators.
>
> Lacking such scrutiny, the organic industry appears to have adopted "black marketing" against conventionally grown foods as its core strategy. The Natural Marketing Institute admitted as much when it reported that "the safety message is a clear driver" of organic sales. A marketing executive for a major organic company was little blunter: "You can, and perhaps should, lead with fear as an industry."
>
> The industry does, in fact, lead with fear. The websites, social media, product packaging, marketing materials and annual reports of organic food companies are full of fear-based advertising against conventional farming. Even more hysterical claims about conventional foods are pushed in food scare campaigns run by NGOs funded by the organic foods industry, as well as by allied natural food and health companies.
>
> In the midst of such claims, where do consumers turn for reliable information? They trust federal regulators to give them the straight scoop based

on science. Yet even here, the federal government is passively complicit in allowing unscientific claims to mislead consumers. Exhibit A in federal complicity is the U.S. Department of Agriculture (USDA) certified Organic label.

USDA's research shows that more than 70 percent of consumers are likely to believe a food is safer, more nutritious or of higher quality if it bears the organic label. In fact, all the label signifies is that a given food has been grown, handled and processed without many of the modern techniques of conventional agriculture.

The label does not even mean that a certain food was grown without pesticides. Organic foods are routinely produced with certain kinds of "organic" pesticides. Meanwhile, organic recalls due to bacterial contamination are ballooning along with the expanding market for organic food.

In short, the federal government is strict about science, labeling and claims for all industries except one. The marketers of organic food are allowed to make scientifically false and misleading claims about the safety and wholesomeness of conventional food, while their products are increasingly likely to be recalled for safety reasons.

Federal agencies have a statutory responsibility to crack down on untruthful and misleading claims in food marketing. They also have a responsibility to warn consumers about real dangers.

The findings by Academics Review raise a number of questions federal regulators should have to answer.

1. Will the USDA, FDA, and Federal Trade Commission enforce existing rules against misleading advertising when marketers misuse the organic label to vilify competitors?
2. Will regulators regard the sponsored attacks on conventional agriculture as advertising, subject to standards of truth?
3. Will the Centers for Disease Control and the FDA investigate what is behind the frighteningly high recall record of organic food?
4. Will the government perform more research on the safety of organic foods?

This is no longer a matter of who wins at the checkout counter. For many vulnerable people, it is a matter of safety. They just don't know it yet.

In 2015, Hank Campbell of the ACSH, in the Science 2.0 report "Organic Food Recalls Up 700 Percent Since 2013,"[25] stated:

Data from USDA and the FDA show that organic food products have accounted for 7 percent of all food units recalled so far this year, compared to 1 percent in 2013. Salmonella, listeria and hepatitis A are not exclusive to organic food, just a lot more common. When you use manure for fertilizer and dupe your customers into thinking you use no pesticides, so the food needn't be washed, they are going to consume feces.

The biggest misconception for organic food's purported increased safety is related to the pesticide issue. This misconception is related to the phrase "pesticides," and the purported dangers associated with chemicals which fall into this category. Further, these misconceptions rely on the typical consumer misunderstanding that organic farming does not use pesticides, which they do. The pesticides available to organic growers fall into the same various categories of risk as conventional ones do: *Practically Non-toxic*, *Slightly Toxic*, and *Moderately Toxic*. Just because the pesticide is from a natural source does not mean it is any less toxic. Steve Savage, Ph.D. (plant pathology), runs the web site Applied Mythology. On September 21, 2012, for Science 2.0, he posted "Pesticides: Probably Less Scary Than You Imagine." In this article, he points out the following:[26]

> The word "pesticide" conjures up negative, scary images. These images come from old organophosphate insecticides of the 1960s that killed fish and birds and caused farm worker illness. These are sorely outdated images. What most people don't know is how much safer the new generations of pesticides are. In fact, scores of old materials have been withdrawn from the market or banned long ago. The new products are mostly compounds with extremely low mammalian toxicity and benign environmental profiles. Today's pesticides are not your grandfather's or even your father's pesticides.

He continues:

> Only 0.2% of the commonly used chemicals in California fall into the 'Highly Toxic' category, and those are used under strict limits to prevent any form of unwanted exposure. Just for interest sake, however, Vitamin D3 would fall into this category if it were a pesticide.

So how does the average consumer make sense of the various EPA toxicity levels in a way that they can apply it to their own lives? Dr. Savage provides the following examples to help illustrate the safety of most commonly used pesticides:

> Vitamin C is something which many people take in large, 250–1000 mg doses on a regular basis. Fifty-five percent (55%) of the pesticides used in California in 2010 were less toxic than Vitamin C. Sixty-four percent (64%) were less toxic than vitamin A. Seventy-one percent (71%) were less toxic than the vanillin in ice cream or lattes. Seventy-six percent (76%) of

the pesticides were less toxic than Prozac and 89% were less toxic than the ibuprofen in products like Advil. Ninety-seven percent (97%) of California pesticide use in 2010 was with products that are less toxic than the caffeine in our daily coffee, the aspirin many take regularly, or the capsaicin in hot sauces or curries. This is not the sort of image that most people visualize when they hear the word pesticides."

These examples leave many consumers confused due to their preconceived prejudice when they hear the word pesticide, and the inherent dangers many of the older pesticides were associated with. However, the newer generation of pesticides have changed dramatically since the 1960s. In "Here's What Electricity Can Teach Us About Pesticide Safety," Dr. Savage wrote for the *Genetic Literacy Project* in 2018:[27]

To understand how something that is designed to kill or otherwise control a pest could be non-hazardous, consider the example of chocolate which has a flavor ingredient that we humans love but which can be toxic to our pet dogs. Chemicals can have different effects on different species. Scientists use the terms specificity and mode of action to describe how chemicals have their specific effects. With modern pesticides, the mode of action is normally the inhibition of some specific enzyme that is important to the viability of the pest. If the enzyme is inhibited by the pesticide, the pest might stop eating, stop growing and/or die.

That enzyme often isn't one that even exists in humans and other animals ourselves or in other groups of organisms unlike the pest. A modern insecticide usually only affects enzymes that are found in insects or even a few kinds of insects. A modern herbicide might only inhibit an enzyme that is needed for the growth of plants. A modern fungicide inhibits an enzyme in a pathway of enzymes that is found in certain fungi. While all of these products should still be handled with a reasonable degree of caution, they are, like the electricity that powers our cell phones, low hazard and thus low risk. We can feel safe about their use.

However, most consumers immediately visualize skull and crossbones when they hear the word "pesticide." However, is this a valid association? If it were, then you literally must stop eating, regardless of the method of production. Why? Because literally all foods contain pesticides, and 99.99% of the pesticides you are exposed to are all natural. These natural pesticides can be equally carcinogenic using the same criteria to judge their synthetic equivalents that the average consumer has learned to demonize.

Bruce Ames, Ph.D., is a molecular geneticist and professor of biochemistry and molecular biology at University of California at Berkeley. He is internationally known for his work on the link between nutrition and DNA integrity, a senior scientist at Children's Hospital Oakland Research Institute, and one of the most widely cited scientific experts with more than 450 publications in leading journals. Dr. Ames points out the following in *Handbook of Pesticide Toxicology* chapter 38, "Pesticide Residue in Food and Cancer Risk":[28]

> We estimate that about 99.9% of the chemicals that humans ingest are naturally occurring. The amounts of synthetic pesticide residues in plant foods are low in comparison to the amount of natural pesticides produced by plants themselves. Of all dietary pesticides that Americans eat, 99.99% are natural: They are the chemicals produced by plants to defend themselves against fungi, insects, and other animal predators. Each plant produces a different array of such chemicals.
>
> We estimate that the daily average U.S. exposure to natural pesticides in the diet is about 1500 mg. . . . In comparison, the total daily exposure to all synthetic pesticide residues combined is about 0.09 mg based on the sum of residues reported by the U.S. Food and Drug Administration (FDA) in its study of the 200 synthetic pesticide residues thought to be of greatest concern. Humans ingest roughly 5000–10,000 different natural pesticides and their breakdown products.
>
> Concentrations of natural pesticides in plants are usually found at parts per thousand or million rather than parts per billion, which is the usual concentration of synthetic pesticide residues. Therefore, because humans are exposed to so many more natural than synthetic chemicals (by weight and by number), human exposure to natural rodent carcinogens, as defined by high-dose rodent tests, is ubiquitous. It is probable that almost every fruit and vegetable in the supermarket contains natural pesticides that are rodent carcinogens. Even though only a tiny proportion of natural pesticides have been tested for carcinogenicity, 37 of 71 that have been tested are rodent carcinogens that are present in the common foods listed in Table 38.1.

Table 38.1 from the text, "Carcinogenicity Status of Natural Pesticides Tested in Rodents," provides the following 37 known natural rodent carcinogens:

> Acetaldehyde methylformylhydrazone, allyl isothiocyanate, arecoline·HCl, benzaldehyde, benzyl acetate, caffeic acid, capsaicin, catechol, clivorine, coumarin, crotonaldehyde, 3,4-dihydrocoumarin, estragole, ethyl acrylate, N_2-γ-glutamyl-p-hydrazinobenzoic acid, hexanal methylformylhydrazine, p-hydrazinobenzoic acid·HCl, hydroquinone, 1-hydroxyanthraquinone, lasiocarpine, d-limonene, 3-methoxycatechol, 8-methoxypsoralen, N-methyl-N-formylhydrazine, α-methylbenzyl alcohol, 3-methylbutanal methylformylhydrazone, 4-methylcatechol, methylhydrazine, monocrotaline, pentanal

methylformylhydrazone, petasitenine, quercetin, reserpine, safrole, senkirkine, sesamol, symphytine.

These naturally occurring toxins are present to fend off diseases, insects, mold, etc. They are simply a natural defensive mechanism of the plant itself to environmental conditions. Table 38.1 also provides examples of the foods where these naturally occurring carcinogens are found, regardless if they are grown organically or conventionally. These foods are:

Allspice, anise, apple, apricot, banana, basil, beet, black pepper, broccoli, Brussels sprouts, cabbage, cantaloupe, caraway, cardamom, carrot, cauliflower, celery, cherries, chili pepper, chocolate, cinnamon, cloves, coffee, collard greens, comfrey herb tea, coriander, corn, currants, dill, eggplant, endive, fennel, garlic, grapefruit, grapes, guava, honey, honeydew melon, horseradish, kale, lemon, lentils, lettuce, licorice, lime, mace, mango, marjoram, mint, mushrooms, mustard, nutmeg, onion, orange, paprika, parsley, parsnip, peach, pear, peas, pineapple, plum, potato, radish, raspberries, rhubarb, rosemary, rutabaga, sage, savory, sesame seeds, soybean, star anise, tarragon, tea, thyme, tomato, turmeric, and turnip.

Dr. Ames also points out that there are "at least 10mg of rodent carcinogens per cup of coffee." But again, don't panic, and enjoy your coffee. As explained earlier, the dosage level you are exposed to is of no threat to your health.

Consider the naturally occurring carcinogens in Table 4.1.

TABLE 4.1 The American Council on Science and Health's "Holiday Dinner Menu"

Food Item	Possible Natural Chemical Carcinogen
1. Cream of mushroom soup	hydrazines
2. Broccoli spears	allyl isothiocyanate
3. Carrots	aniline, caffeic acid
4. Baked potato	ethyl alcohol, caffeic acid
5. Cherry tomatoes	benzaldehyde, hydrogen peroxide, quercetin glycosides
6. Sweet potato	ethyl alcohol, furfural
7. Celery	caffeic acid, furan derivatives, psoralens
8. Assorted nuts	aflatoxin, furfural
9. Pumpkin pie	benzo(a)pyrene, coumarin, safrole
10. Roast turkey	heterocyclic amines
11. Red wine	ethyl alcohol, ethyl carbonate
12. Comfrey tea	symphytine
13. Jasmine tea	benzyl acetate
14. Coffee	benzo(a)pyrene, benzaldehyde, benzene, benzofuran, caffeic acid, etc. (coffee contains 14 potential carcinogens)

Clearly, millions of people consume these food items listed on the "Holiday Dinner Menu" and their naturally occurring chemicals each day, as well as those listed by Dr. Ames, and remain perfectly healthy. If these natural toxins which occur in higher concentrations than synthetic pesticides are equally carcinogenic, then why do the media and environmentalists ignore this "health hazard"? Simple: because if you inform consumers that their exposure to the synthetic chemicals used to grow their food is infinitesimal to non-existent by the time they consume them and therefore no real hazard exists (just a theoretical "The Boy Who Cried Wolf" situation), then you are unable to create the separate, fabricated market for your products—organics. To tell the truth would essentially destroy what has been deceptively created.

A main issue consumers need to understand regarding the potential health hazards associated with any chemical is the *Principle of Toxicology—the dose makes the poison*, which has been redundantly stated throughout this book. All chemicals, initially perceived to be either good or bad for you, fall under this principle. The essential point here is that no chemical compound is inherently bad for you until your level of exposure surpasses its upper level of safety. This principle also applies to the naturally occurring rodent carcinogens mentioned previously. One or several cups of coffee per day is not harmful due to this principle.

These illustrations of the myth that "natural" inherently means "safe" can be found in any college nutrition text. All one needs to do is pick up any beginning college nutrition text and review the tolerable upper intake levels for most vitamins and minerals. You will notice "chemicals," such as vitamins A, D, C, B6, etc., as well as minerals iodine, iron, fluoride, arsenic, etc., are all examples of naturally occurring chemicals which are safe at low doses, but hazardous and often deadly at higher dosages. Just because they are "natural" does not mean they are inherently always safe.

But this issue becomes a moot point when consumers understand how very little, if any, pesticide residue they are ever even exposed to in the first place by the time the produce reaches them.

The most recent national FDA Pesticide Monitoring report released in 2018,[29] after analyzing 7,413 produce samples, states that "no pesticide chemical residues were found in 52.9% of all domestic and 50.7% of import samples analyzed," which is consistent with past years. For the state of California, which, according to the USDA, is the top agricultural producing state in the United States, the California Department of Pesticides Regulation (DPR) report "Pesticide Info—What You Should Know About Pesticides: Pesticides and Food: How We Test for Safety," states "no residues are detected in about 60%" of crops.

On November 7, 2012 the DPR published "DPR 2011 Monitoring Shows Most Produce Samples Have No Detectable Pesticide Residues," after testing 2,707 samples of more than 160 types of domestic and imported produce.[30] In this report, Director Brian R. Leahy stated, "we want to emphasize that most produce has no detectable pesticide residues and when there are residues, they are at such a low level they are not a health risk." Further, if residues were found, over 99% of domestic and 90% of import human foods were compliant with federal standards, which are set at such low levels the margin of safety is significant. The standard for pesticide safety is a maximum residue at least 10–100 times below the exposure found *not to cause* [my emphasis] adverse effects in test animals. This is an incredibly liberal safety net, and likely unnecessary. Pesticides are occasionally present in our food to be sure. Yet, as I have shown, once placed in perspective, the amount is minute and irrelevant. Be thankful that we have them available to us to use, unless you wish to return to an era of producing your daily food as well as paying considerably more for it.

To further illustrate the minimal exposure, I contacted a local, large farming operation in Bakersfield, California, Sun-World International Inc., and simply asked for some independent lab results of their pesticide residue testing program. David Fenn, at the time a Senior VP who managed food safety and inspections, forwarded the pesticide residue analysis results from Primus Labs in Santa Maria, California for their mandarins. The samples were taken November 17, 2007 for the chemicals methyl carbamates, organohalides, and organophosphates, and the results were: none detected in ppm.

Per David Fenn in a March 17, 2008 email, "we normally see no detectable residues on our crops." This is consistent with both the USDA national testing program for conventionally grown crops and the independent testing of the California Department of Agriculture mentioned earlier. The point the reader should take away from David Fenn's comment, as well as the USDA national pesticide testing program and the California pesticide testing program, should be obvious. If you are buying organic because you believe they are safer for you due to some purported pesticide residue, then you have been deceived. Most conventionally grown produce by the time you purchase it at the store will have no pesticide residue, and those that do are so minimal they should be of no concern to you.

To illustrate further the minuscule, if any, exposure consumers have to pesticides, consider the following illustrations provided by the American Crop Protection Association (ACPA) in their publication *Growing Possibilities*. In this publication, the ACPA points out that the following examples "refers to different, single, commonly used pesticide on a popular food crop."

- A 150-pound adult could eat 3,000 heads of lettuce every day for a lifetime and not exceed the levels of pesticide residue that have been proven to have no effect on laboratory animals.
- A baby could be fed 87 cups of applesauce every day.
- A 40-pound child could munch 30,000 pounds of carrots every day.
- A 132-pound adult could eat 396,000 pounds of bananas daily.

For a number of other examples, the reader is encouraged to go to www. safefruitsandveggies.com and follow the links for the pesticide residue calculator.

This program will allow you select adult male or female, teen, or child and then an option for 19 different fruits or vegetables. After selecting the produce of choice, the site will then provide you with how many servings you could consume per day without any effect from the pesticide used in that commodity even if the commodity had the highest pesticide residue recorded by the USDA.

As an example, if you choose the adult male option and then strawberries, the servings of strawberries an adult male could consume daily is 635 servings, even if the strawberries had the highest residues allowed by the USDA.

Pesticides, due to the instability of most them, are normally applied in higher concentrations during the crop cycle when the plant is most likely to be invaded by pest species, and possibly at lower concentrations just prior to harvesting. Since most modern pesticides are very unstable when exposed to sunlight, water, other elements, and microorganisms, they degrade very quickly and leave negligible residue on the crops by the time they are either harvested or reach the market.

However, there are certainly those pesticides that are resistant to degradation by light, water, heat, air, etc., and can accumulate in both the soil and the food chain. These pesticides are highly regulated, and, as with all pesticides, acute toxicity is generally due to accidental spills or possibly by mishandling the pesticide by those applying the pesticide in the field. These occurrences, while possible, are rare, and the acute toxicity that results is not due to residues left on the food. To assist in risk control, many pesticides can only be applied by certified licensed applicators whose livelihood relies on following the guidelines required by law. Also, keep in mind that the use of pesticides can be a significant expense to the farmer. This expense gives them the incentive to use such products judiciously and wisely about both timing and quantity. Additionally, if the farmer's product is found to contain levels above what the EPA has set for human exposure, the commodity is subject to seizure. This is obviously a very costly mistake for any grower, and one that provides enormous incentive to stay within the established safe standards.

Recently, I debated this issue with an individual who was an avid organic enthusiast due to the "potential carcinogens" in the food supply. However, I found it ironic that he did not seem concerned about his obesity (obesity, sedentary lifestyle, and smoking accounts for most of the cancers), nor the cup of coffee he was sipping on that contains over a dozen natural occurring cancer-causing chemicals when given to rodents in large doses (see the "Holiday Dinner Menu" presented earlier). He was also unaware that he gets more carcinogens in that one cup of coffee than pesticide residue in a year. As an example, in 2014, the *Journal of Agriculture and Food Chemistry* stated, "our findings suggest that caffeic acid (in coffee) can be implicated as a potent insecticidal molecule and explored for the development of effective dietary pesticide."[31]

All growers I have spoken to over the years declare the same frustration over the rampant misguided perspectives on pesticides. All essentially agree with the following statement made by one grower: "When will people start looking at pesticides and fungicides as they look at modern medicine and that they (pesticides) have improved our lives and health?"

Finally, keep in mind that technology now allows farming operations to boost crop yields and minimize pesticide use by using satellite maps and computers to match applications to soil and crop conditions as well as integrated pest management. Pesticides are simply one of the tools we need to eat.

To help you keep a proper perspective regarding reports on pesticide residues in food and water, consumers need to understand that just because a pesticide or any chemical has been detected on food, this does not equate to harm, even though many "consumer" advocacy and environmental groups would like you to believe it does. Modern analytical or detection instruments can detect the even infinitesimal and harmless chemical residues which are expressed as parts per million (ppm), parts per billion (ppb), or parts per trillion (ppt). The following comparisons may help put these quantities into perspective.[32]

- **1 ppm** = 1 gram of residue in 1,000,000 g of food; 1 inch in 16 miles; 1 minute in 2 years; 1 cent in $10,000; or 1 pancake in a stack 4 miles high.
- **1 ppb** = 1 gram of residue in 1,000,000,000 g of food; 1 inch in 16,000 miles; 1 second in 32 years; or 1 cent in $10 million.
- **1 ppt** = 1 gram of residue in 1,000,000,000,000 g of food; 1 inch in 16 million miles; 1 second in 32,000 years; 1 square foot of floor tile on a floor the size of the state of Indiana.

Crop Life America provides these examples:[33]

- One part per million = only ten bricks of the Empire State Building. At this level of exposure, a child could eat 5,291 servings of blueberries in one day without any harm from pesticides even if the berries had the highest pesticide residue recorded by the USDA. This serving is approximately 860 pounds, equivalent to the weight of a horse!
- The book *Harry Potter and the Order of the Phoenix* contains a total of 257,045 words. If you had a stack of 3,890 of those books, one word in one of those books = 1 part per billion. Pesticide exposure at this level would allow a man to eat 25,339 servings of carrots in one day without any harm from pesticide effect even if they had the highest pesticide residue recorded for carrots by USDA. This serving is approximately 12,669 pounds, equivalent to the weight of an elephant!

Naturally occurring toxins, approximately 10,000 of them, may be present in parts per thousand or millionth (higher concentrations than synthetics), whereas synthetic residues are found in foods at parts per billions or lower. In 2018, the population of China surpassed 1.4 billion. So, to provide another visual of just how small 1 part per billion really is, think of it as less than one person in all of China.

In summary of this section, compare the following statement made by the California Department of Pesticide Regulation report "Pesticides and Food: How We Test for Safety"[34] to what I illustrated in the liability of supplements chapter.

> Pesticides are among the most regulated products in the country. Before a pesticide can be used in California, it must be evaluated and licensed by both the EPA and the California Department of Pesticide Regulation. The manufacturer must present test data to show the pesticide will not pose unacceptable risks to workers, consumers, or the environment.

This may take as long as 11 years and extensive expense, according to Crop Life America. In contrast, supplements, which over half of the US population consumes, go through zero mandatory premarketing evaluation for safety and efficacy, and send thousands of consumers to local emergency rooms per year (see Chapter 9). This is why I believe a consumer's minuscule to no exposure to pesticides is far safer than their exposure to many supplements. Ironically, both the supplement industry and the organic food industry have learned to use the same successful negative narrative marketing method to establish their fabricated markets—fear.

Misunderstanding 3: Does Organic Food Taste Better? The Halo Effect

So, if organic food is not necessarily safer or healthier, does it at least taste better? Often, a consumer will attempt to compare the taste of a locally grown, fresh organic product with its national chain-store counterpart. This is not a fair comparison. Any product grown locally, organic or not, will have a better taste to it vs. one that was harvested and shipped from elsewhere, regardless of its growing methods. A tomato grown in your backyard, under either condition (organically or with the use of pesticides) will have the same flavor. The key regarding taste is not whether it is conventionally grown or organic; it is how fresh or ripe the produce was when picked and made available to you.

Many consumers are victimized or deceived by their own expectations of the word "organic." Due to the many misunderstandings, most consumers are simply victimized by the "health halo effect" of the word organic. As an example, on April 26, 2011, researchers at Cornell University reported the following in the *Cornell Chronicle* report: "Organic food labels create perception of healthier fare."[35] When 115 volunteers had to rate the taste of a set of three different foods, all organically grown, but with one pair labeled conventionally grown while the other was labeled organic, most volunteers not only felt those foods labeled organic tasted better, but were also lower in calories and nutritionally superior. In reality, however, every food item they tasted were the same. The "organic" label made the difference in the consumer's perception of the quality and taste of the food item. This deceptive "halo effect" of organics is the foundation of the organic industry. It is not science; it is simply the epitome of the deceptive marketing of organics on misinformed consumers.

On October 20, 2014, two Dutch pranksters pulled off a similar event when they recorded their attendance at an organic food expo promoting their "new organic alternative to fast food." They purchased some classic McDonald's food, reconfigured the packaging of it to appear "natural," and then served it to several "experts." The responses were exactly what one would expect from the placebo, health halo effect of the word organic. The Dutch crew stated at the end, "*if you tell people that something is organic, they'll automatically believe it's organic!*" The video can be found at www.youtube.com/watch?v=4Qa6QXBxxWw.

Misunderstanding 4: Food Production and Environmental Benefits

The average consumer does not understand the enormous benefits in yield that is received from pesticides. On August 23, 2016, the journal *PLOS/One* published its findings comparing yields per acre of organics vs. conventionally grown. The researchers state:

> The analysis we present here offers a new perspective, based on organic yield data collected from over 10,000 organic farmers representing nearly 800,000 hectares of organic farmland. We used publicly available data from the United States Department of Agriculture to estimate yield differences between organic and conventional production methods for the 2014 production year. Similar to previous work, organic crop yields in our analysis were lower than conventional crop yields for most crops. Averaged across all crops, organic yield averaged 80% of conventional yield.[36]
> Some yield deficits, potatoes for example, were as high as 60%.

So, given this information, here is a theoretical example I give in my nutrition class to illustrate a simple point to students so that they evaluate the widespread belief that organic agriculture is more environmentally friendly. You are provided 2,000 acres of farmland to feed 5,000 people, and if you choose to use conventional farming methods, the yield you produce will meet the required need. However, you decide to go the organic route for your production process. At the end of the growing season, you have come up short of the necessary yield by 20%. So, in order for you to meet your necessary yield target to feed those 5,000 individuals the next year, you are going to have to bring into production at least 20% more acreage (which is at least 400 more acres, but likely more), 20% more water use, 20% more fuel use, etc. Now multiply this on a worldwide scale to feed over 7 billion people, not just my theoretical 5,000, and you can begin to grasp the benefits of conventional agricultural practices with the use of safe and effective pesticides.

Steve Savage, Ph.D., provided an illustration of this in his article "The Lower Productivity of Organic Farming: A New Analysis and Its Big Implications."[37] He reviewed the 2014 USDA detailed survey of organic crops and stated, "to have raised all U.S. crops as organic in 2014 would have required farming one hundred nine million more acres of land."[38] In his review, "The Organic Yield Gap," he stated the following in the High Level Summary section:

> The productivity of organic agriculture tends to be lower than that of conventional. That is part of why there is a need for a price premium to the grower. Some have argued that this yield gap can be closed, but the data from the latest, detailed, 2014 USDA organic survey suggests that real world organic yields are substantially lower. This analysis is based on 371 crop/geography comparisons representing 80% of all US cropland. In 84% of the crop/geography comparisons, organic yields were lower-mostly in the 20–50% range. In order for the U.S. crop production from 2014 to have been produced as organic would have required at least 109 million more farmed acres—an area equivalent to the total parkland and wild lands in the lower 48 states. Organic remains a very small fraction of the US cropland base (~0.44%) and so it puts a limited strain on land use. That said, the concept of 'only organic' is untenable from an environmental perspective.[39]

In March of 2017, I had a conversation with a local grape grower here in Kern County who grows both organic as well as conventionally grown grapes. He stated that to control weeds in his organic vineyards, he must spend roughly $500 per acre due to the labor involved in hand hoeing vs. $20 per acre with herbicides for his conventional crop. He had already spent thousands of dollars for just weed control in his organic vineyards, and he stated that if he sprays with what is allowed for organic crops, he must spray 2–3 times more, due to the lack of effectiveness of the organic herbicide allowed.

He also stated that the previous year, 40 acres of his organic grapes became infected with mites. His options were to either allow the crop to go bad and lose the produce, or spray the crop to salvage the grapes. Spraying would result in losing the organic certification for three years for that acreage as a result, but he believed that it was the obvious choice to salvage the produce. Ironically, he also stated something which is true for many conventionally grown crops that are managed well: even his conventionally grown grapes never have any pesticide residue on them when shipped to market. The point here is that even though he sprayed his "organic" crop to destroy the mites, the level of pesticide residue on the end product was zero, which is the same as his organic varieties grown in other fields. So, at the marketplace, there would be no distinguishable difference regarding pesticide residue between his "organic" variety and his conventionally grown variety, just the higher cost of the organics to the misinformed consumer.

Finally, a German study published in the *Journal of Cleaner Production* on September 10, 2017, "Carbon Footprints and Land Use of Conventional and Organic Diets in Germany," found:[40]

1. The carbon footprints of the average conventional and organic diets are essentially equal.
2. The average organic diet uses 40% more land than the average conventional diet.

So, this begs the question: which type of agricultural practice is more environmentally friendly, organic or conventional? I would argue that conventional agricultural practices that allow for the introduction of advances in chemical, mechanical, and genetic modification for better pest control, increased yield per acre, and reduction in use of pesticide, fuel, and water, etc., is far more environmentally and consumer-friendly than organic could ever be.

An excellent and more extended summary of this issue can be found in "Is Organic Farming Better for the Environment?" published in 2017 by Steve Savage, Ph.D.[41]

The highlights of Dr. Savage's review are as follows:

- Organic farmers are more dependent on older, "natural," less targeted chemical pesticides that can be more toxic and harm beneficial insects.
- Organic's 15–50% yield gap means expansion of organics pressures limited land resources with negative environmental impact.
- Organic rules block farmers from using state-of-the-art soil building practices.
- Genetic engineering encouraged wider adoption of ecologically protective no-till farming.
- Farm sustainability is best promoted by using best practices, regardless whether organic or conventional.

Misunderstanding 5: Is It Even Organic in the First Place?

In September 2017, the USDA published its audit report of the National Organic Program—International Trade Arrangements and Agreements from the Office of Inspector General.[42] In this report, after the USDA audited seven U.S. ports of entry, they concluded the following:

1. "The National Organic Program was unable to provide reasonable assurance that the required documents were reviewed at U.S. ports of entry to verify that imported agricultural products labeled as organic were from certified organic foreign farms and businesses that produce and sell organic products."[43]
2. "Imported agricultural products, whether organic or conventional, are sometimes fumigated at U.S. ports of entry to prevent prohibited pests from

entering the United States." And, there is no "established and implemented controls at U.S. ports of entry to identify, track, and ensure treated organic products are not sold, labeled, or represented as organic."[44]

In May of 2017, Peter Whoriskey, an investigative reporter for the *Washington Post*, tracked the shipments of purported organic corn and soybeans from overseas sources to US ports of entry. He detailed the finding of the following major points:[45]

1. A total of 36 million pounds of soybeans sailed from the Ukraine to Turkey to California. Along the way, it underwent a remarkable transformation. The cargo began as ordinary soybeans, according to the documents obtained by the *Washington Post*, fumigated with pesticides and priced liked ordinary soybeans. By the time they reached California, the soybeans had been labeled "organic," according to receipts, invoices and other shipping records.
2. The multimillion-dollar metamorphosis of the soybeans, as well as two other similar shipments in the past year examined by the *Post*, demonstrate weaknesses in the way that the United States ensures that what is sold as "USDA Organic" is really organic.
3. The imported corn and soybean shipments examined by the *Post* were largely "destined to become animal feed and enter the supply chain for some of the largest organic food industries. Organic eggs, organic milk, organic chicken and organic beef are supposed to come from animals that consume organic feed."

The December 3, 2012 issue of *Delta Farm Press* made some interesting points in the article "Organic Food Inspections Lacking, Despite USDA Program" by Forrest Laws.[46] In an interview with Mischa Popoff, currently a policy advisor for the Heartland Institute and author of the book *Is It Organic? The Inside Story of Who Destroyed the Organic Industry, Turned It into a Socialist Movement and Made Million$ in the Process*." Mr. Popoff grew up on an organic grain farm and had a prior background as an organic crop inspector. He made the following points:

*   Organic certification is misleading. The checks and balances which prevent most organic growers from using conventional pesticides are simply not in place. This is even though organic farmers are required to file extensive amounts of paperwork to "satisfy the USDA certification program, but this does not require any verification of the claims the crops have not received applications of synthetic fertilizers or been treated with pesticides."[47]
*   The testing standards for organic produce is generally done on end products and not in the field, "so it is really useless because a lot of the substances that are prohibited in organic food production have dissipated by the time they're harvested,"[48] just as in conventional farming. Popoff points out that modern agriculture has done away with persistent chemicals and moved to pesticides

which dissipate rapidly. Therefore, they're safe, and you won't see these in the end products of either organic or conventionally grown produce, which is a major point of this chapter.

To support Popoff's first point, in 2006 California Liquid Fertilizer was caught selling "organic" fertilizer which had been spiked with ammonium sulfate, a synthetic fertilizer not allowed by organic farms.[49] According to the report, California Liquid Fertilizer held as much as one-third of the California organic market, indicating that much of the "organically" harvested produce were not really organic.

On October 12, 2018, the US Department of Justice news release "Three Nebraska Farmers Plead Guilty to Fraud Involving Sales of Grain Fraudulently Marketed as Organic," made the following statement:[50]

> At their respective plea hearings, each man admitted to growing grain between 2010 and 2017 that was not organic. Each further admitted that they knew the grain was being marketed and sold as organic, even though it was not in fact organically grown. The charging documents allege that, during the 2010 to 2017 period, each of the three farmers received more than $2.5 million for grain marketed as organic.

Again, as pointed out by the *Post* investigative report discussed previously, it must be wondered how much of this "organic" grain over that seven-year period was "destined to become animal feed and enter the supply chain for some of the largest organic food industries. Organic eggs, organic milk, organic chicken and organic beef are supposed to come from animals that consume organic feed."

On May 10, 2019, the Genetic Literacy Project published "Why the 'Chemical Free' Organic Industry Has a Pesticide Problem." In this report, they state:[51]

> Several years ago (2014) a Canadian study sent shockwaves through the agricultural community, when the government discovered that more than half of all labeled as "organic" instead had pesticide residues. The government insisted that the amounts were not life-threatening, but organic foods aren't supposed to have any pesticide residues (outside of those allowed on organic crops). None. Instead, the study found that 77 percent of organic grapes, 67 percent of organic strawberries and 59 percent of organic tomatoes contained pesticides. This wasn't a domestic production problem, but an import issue; four-fifths of Canada's produce is imported.[52]

Closing Comments

Without the use of pesticides, the poor and the middle class would be far less likely to be able to afford the quality and quantity of fruits and vegetables necessary for good health. Most of the pesticide phobia has been a result of false reliance on extrapolating large-dose animal research to humans. This "mouse as a little man" premise is a very shaky proposition due to the distinct differences in animal and human physiology and anatomy. Jack Fisher, M.D., at the University of California, San Diego, coined the term "Mouse Terrorism" to fit this extrapolation of data and the erroneous notion that carcinogenicity in humans is inferable from data on the risk of cancer in man-made species of rodents.

There is no question that the benefits of a balanced diet rich in fruits and vegetables, from either organically or conventionally produced agricultural practices, far outweigh the largely theoretical risks posed by occasional very low to no pesticide residue levels in foods, again, from either organic or conventionally grown foods.

So instead of needlessly worrying about some theoretical cancer or health risk purportedly associated with synthetic chemicals, which is unfounded, consumers should focus on the real risks they are so negligent about, such as tobacco use, obesity, sedentary lifestyle, poor diets, and excess alcohol and drug use. These issues account for most of the projected $4.6 trillion we are expected to spend on healthcare by 2020 and are the prevailing burdens of any society.[53] So, find another phobia. Chemicals and pesticides are not the problem; an unhealthy lifestyle is— and it is literally and financially killing us.

Now, referring to the beginning of this chapter, my hope is that:

- Your fear of the responsible use of chemicals has now passed.
- Be thankful that the technology to grow our food has advanced so that 1–2% of the population can grow what the 100% needs.
- Stop needlessly worrying about hypothetical fears but instead concern yourself with real issues in your life.

Hopefully, you now understand that:

- A pesticide is a chemical, nothing more, and as with all chemicals, including vitamins and minerals, water, etc., it is the dose that makes the poison not the chemical—*the Principle of Toxicology*.
- Chemicals allow us to feed a growing population with increasingly less agricultural land to produce it on.
- Chemicals allow for far more people to consume healthy foods which will diminish their cancer risks vs. increasing it.
- Chemicals are an integral part of the safe and productive food supply process.
- The produce's vitamin content you choose to consume is related to the genetics of the plant, and the mineral content is related to the composition of the soil it was grown in, regardless if it was conventionally grown or organically grown.
- Pesticides and related chemicals are an integral part of being responsible stewards of the Earth.

Notes

1. F. Todd, Pistachios push Kern County crop values to No. 1, *Western Farm Press* (September 19, 2017), p.1. farmprogress.com/crops/pistachios-push-kern-county-crop-values-no-1
2. J.J. Cohrssen and H.I. Miller, The USDA's meaningless organic label. *Agriculture* (Spring 2016), p. 24.
3. B. Chassy, Ph.D. Why consumers pay more for organic foods? Fear sells and marketers know it. *Academic Reviews* (April 8, 2014). academicsreview.org/2014/04/why-consumers-pay-more-for-organic-foods-fear-sells-and-marketers-know-it/
4. Ibid, p. 1.
5. J. Block, Former U.S. Secretary of Agriculture Glickman criticizes organic industry for misleading marketing. Genetic Literacy Project-Science Not Ideology (May 16, 2014). geneticliteracyproject.org/2014/05/16/former-us-secretary-of-agriculture-glickman-criticizes-organic-industry-for-misleading-marketing/
6. Ibid, p. 1.
7. Ibid, p. 1
8. USDA, *Agricultural Resources and Environmental Indicators*, chapter 4.3, p.1 ers.usda.gov/webdocs/publications/41964/30294_pestmgt.pdf?v=41143
9. USDA National Institute of Food and Agriculture. *Global Scientists Meet for Integrated Pest Management Idea Sharing.* April 14, 2015. p. 1.
10. J. Stossel, Besieging Stossel to protect organic myths. *Priorities for Health* (February 4, 2000), Vol. 12, No. 3, pp. 7–10.
11. Ibid, p. 1.
12. pennstatefoodsafety.blogspot.com/2014/10/organice-food-safety-fact-versus-hype.html
13. C. Smith-Spangler, M.L. Brandeau, G.E. Hunter, et al. Are organic foods safer or healthier than conventional alternatives?: A systematic review. *Annals of Internal Medicine* (2012), Vol. 157, p. 348–66.
14. A.D. Dangour, S.K. Dodhia, A. Hayter, E. Allen, K. Lock, R. Uauy, Nutritional quality of organic foods: A systematic review. *The American Journal of Clinical Nutrition* (September 2009), Vol. 90, No. 3, pp. 680–85.
15. M. Kristensen, L.F. Østergaard, U. Halekoh, H. Jørgensen, C. Lauridsen, K. Brandt, and S. Bügel, Effect of plant cultivation methods on content of major and trace elements in foodstuffs and retention in rats. *Journal of the Science of Food and Agriculture* (2008), Vol. 88, pp. 2161–72. https://onlinelibrary.wiley.com/doi/abs/10.1002/jsfa.3328
16. Ibid, p. 1.
17. C.S. Williamson, Is organic food better for our health? *Nutrition Bulletin* (June 2007), Vol. 32, No. 2, pp. 104–8.
18. How much are pesticides hurting your health? *Tufts University Diet and Nutrition Letter* (April 1996), Vol. 14, No. 2, p. 4.
19. gmo.geneticliteracyproject.org/FAQ/are-organic-foods-healthier-than-conventional-foods/
20. foodsafetynews.com/2012/03/an-outbreak-like-germanys-could-happen-here/
21. Ibid, p. 2.
22. F. Pan, X. Li, J. Carabez, G. Ragosta, K. L. Fernandez, E. Want, A. Thiptara, E. Antaki, and E.R. Atwill, Cross-sectional survey of indicator and pathogenic bacteria on vegetables sold from Asian vendors at farmers' markets in Northern California. *Journal of Food Protection* (March 2015), Vol. 78, No. 3, pp. 602–8.
23. https://pennstatefoodsafety.blogspot.com/2014/10/organice-food-safety-fact-versus-hype.html
24. https://www.desmoinesregister.com/story/opinion/columnists/2014/10/07/john-block-consumers-misled-organic-food-safety/16840717/

25. https://www.science20.com/science_20/organic_food_recalls_up_700_percent_since_2013-156919
26. https://www.science20.com/agricultural_realism/pesticides_probably_less_scary_you_imagine-94343
27. https://geneticliteracyproject.org/2018/09/07/electricity-pesticides/
28. B. Ames, Pesticide residues in food and cancer risk: A critical analysis. In *Handbook of Pesticide Toxicology*, 2nd ed., edited by R. Krieger, pp. 800–1. San Diego, CA: Academic Press, 2001.
29. https://www.fda.gov/food/cfsan-constituent-updates/fy-2016-pesticide-analysis-demonstrates-consistent-trends-over-five-years
30. https://www.cdpr.ca.gov/docs/pressrls/2012/121107.htm
31. R.S. Joshi, T.P. Wagh, N. Sharma, F.A. Mulani, U. Sonavane, H.V. Thulasiram, R. Joshi, V. S. Gupta, and A. P. Giri, Way toward "Dietary Pesticides": Molecular investigation of insecticidal action of caffeic acid against "Helicoverpa armigera". *Journal of Agriculture and Food Chemistry* (November 12, 2014), Vol. 62, No. 45, pp. 10847–54.
32. *IFIC review: Pesticides and food safety*. International Food Information Council Foundation (October 15, 2009), p. 2.
33. https://static1.squarespace.com/static/59b55b2b37c581fbf88309c2/t/5c8ba165971a1869c8fe3e2d/1552654696456/Residue+Infographv2.jpg
34. https://www.cdpr.ca.gov/docs/dept/factshts/residu2.pdf
35. Organic food labels create perception of healthier fare. *Cornell Chronicle* (April 26, 2011) news.cornell.edu/stories/2011/04/organic-food-label-imparts-health-halo-study-finds
36. A.R. Kniss, S.D. Savage, R. Jabbour, Correction: Commercial crop yields reveal strengths and weaknesses for organic agriculture in the United States. *PLOS ONE* (2016), Vol. 11, No. 11.
37. S. Savage, The lower productivity of organic farming: A new analysis and its big implications. *Forbes* (October 9, 2015).
38. Ibid, p. 1.
39. scribd.com/doc/283996769/The-Yield-Gap-For-Organic-Farming
40. H. Treu, M. Nordborg, C. Cederberg, T. Heuer, E. Claupein, H. Hoffmann, and G. Berndes, Carbon footprints and land use of conventional and organic diets in Germany. *Journal of Cleaner Production* (2017), Vol. 161, pp. 127–42.
41. https://geneticliteracyproject.org/2017/02/16/organic-farming-better-environment/
42. *USDA Office of Inspector General: National Organic Program – International Trade Arrangements and Agreements.* Audit Report (September 2017) Report No. 01601-0001-21.
43. Ibid, p. 7.
44. Ibid, p. 12.
45. P. Whorisky, The labels said 'organic.' But these massive imports of corn and soybeans weren't. *The Washington Post* (May 12, 2017).
46. Forrest Laws, *Organic food inspections lacking, despite USDA program*. Delta Farm Press (December 3, 2012).
47. Ibid, p. 5.
48. Ibid, p. 5.
49. mercedsunstar.com/news/business/agriculture/article3239011.html
50. Three Nebraska Farmers Plead Guilty to Fraud Involving Sales of Grain Fraudulently Marketed as Organic, United States Department of Justice. Northern District of Iowa (October 12, 2018). justice.gov/usao-ndia/pr/three-nebraska-farmers-plead-guilty-fraud-involving-sales-grain-fraudulently-marketed
51. A. Porterfield, Why the 'chemical free' organic industry has a 'pesticide problem, *Genetic Literacy Project* (May 10, 2019).
52. Ibid, p. 2.
53. P. Galewitz, Nation's Health Care Bill To Nearly Double By 2020. *Kaiser Health News* (July 28, 2011).

5

OBESITY

Whose Responsibility Is It? The Blame Game

No one is going to argue that we have a serious weight control and obesity issue in the United States, as well as almost every industrialized country. It is a simple observation. According to a 2017 report from the Centers for Disease Control, National Center for Health Statistics (NCHS), the prevalence of obesity among adults is 40% and among adolescents from 12–19 years old, it is 20%.[1] Additionally, according to the NCHS report, nearly 71% of Americans are either overweight or obese. A February 13, 2018 report by The Heritage Foundation, "The Looming National Security Crisis: Young Americans Unable to Serve in the Military," states that 75% of Americans ages 17–24 are unable to join the military, with the biggest culprit keeping young adults from qualifying to serve as health concerns related to obesity.

The health care cost for obesity-related issues is already staggering: currently at roughly $200 billion per year, and only progressively getting worse.[2] Further, obesity-related issues are even more problematic than those associated with smoking, considering the tsunami of chronic preventable diseases which occur as a result. According to the CDC, these diseases include the following:[3]

- Coronary artery disease.
- Type 2 diabetes.
- Cancers (endometrial, breast, and colon).
- Hypertension.
- Stroke.
- Dyslipidemia.
- Liver and gallbladder disease.
- Sleep apnea and respiratory problems.
- Osteoarthritis (joint degeneration).

- Gynecological problems (abnormal menses and infertility).
- Etc., possibly as many as 40 different disease conditions.

Images 5.1 and 5.2, from the RAND Corporation's report "The Risks of Obesity Worse Than Smoking, Drinking or Poverty," provides a graphic illustration of the burden obesity is placing on healthcare costs needed to manage chronic conditions associated with obesity.[4]

This estimated cost for the treatment of obesity-related conditions does not even consider that obese employees have a documented increased rate of workers' compensation costs and lost workdays. Additionally, a study conducted by the combined efforts of UCLA and the University of Pittsburgh illustrated that with increased rates of body fat, a direct association with "detectable brain volume deficits" also occurs.[5] Researchers found that obese individuals have 8% less brain tissue than normal-weight individuals as they age. With the *assumption* [my emphasis due to the fact that it is an association only] that this will precipitate higher rates of dementia in this population, the link between loss of brain function and obesity

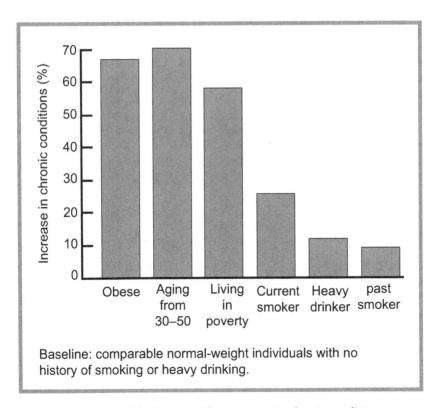

Baseline: comparable normal-weight individuals with no history of smoking or heavy drinking.

IMAGE 5.1   Obesity is linked to a significant increase in chronic conditions.

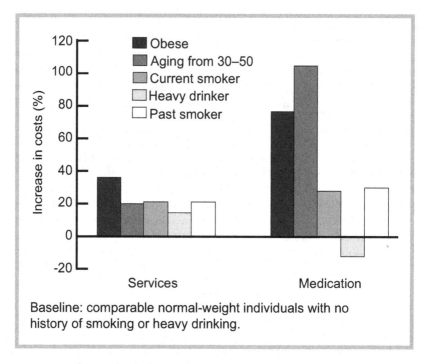

IMAGE 5.2 Obese individuals spend more on health care than smokers and heavy drinkers; only aging has a greater effect on medication costs.

only indicates possible further economic hurdles to overcome as this population ages—if this assumption is accurate.

However, enough of the statistics, projected health care costs, and physical toll obesity is going to have. The bigger questions are: whose fault is it, and what—if anything—can we do about it?

Whose Fault Is It?

The classic response to any problem anymore is "what is the government going to do about it?" This response showcases a burgeoning new American tradition of abdicating our personal responsibility to the federal government, which is apparently, according to the belief of many, supposed to be our nanny in every aspect of our lives. This includes the regulating of our exposure to sodas, Twinkies, and chips, and demonizing the companies who dare manufacture and sell "junk food."

As an example, the very popular *New York Times* writer Jane Brody in her September 12, 2011 article "Attacking the Obesity Epidemic, First Figuring Out Its Cause,"[6] writes that "if you have gained a lot of unwanted pounds at any time during the last 30-odd years, you may be relieved to know that you are probably not to blame. At least not entirely."[7] She goes on to blame "many environmental forces" such as the food and beverage industry to "the way our cities and towns are built."[8] It is simply another article abdicating individual and parental responsibilities and portraying obese individuals as victims vs. being responsible for the choices they or their parents have made. She, as well as many others, believe that government taxes on "unhealthy" foods and beverages would help, but it will not. Who decides what is "unhealthy"—since, essentially, any food item can be determined unhealthy if you consume too much of it?

But abdicating our personal responsibility to the government is not just an American behavior, as the same appears to be occurring in the United Kingdom (UK). In the UK, Public Health England (PHE) is considering "plans to cap the calories in thousands of meals sold in restaurants and supermarkets" to help curb Britain's obesity crisis, according to *The Daily Telegraph*'s October 11, 2018 article "Pizza Must Shrink or Lose Their Toppings Under Government Anti-Obesity Plan."[9] The article also quotes Dr. Alison Tedstone, the chief nutritionist for PHE, as stating, "we know that just having healthy options on the menu won't change the nations habits,"[10] which is precisely the point of this chapter. We have a growing population of people who are failing to take personal responsibility for themselves and their choices, as well as their roles as a parent.

Instead of personal responsibility, we now have many who would like to put a significant portion of the blame for the obesity epidemic on the marketing of foods and beverages that are believed to be associated with it. However, on the first night of a class, I teach my 18–21-year-old beginning nutrition students the distinct difference between an "associated" cause of something and an actual "cause-and-effect" relationship. The advertisement of the food or the readily availability of it does not make you obese; the sedentary time watching it, as well as the irresponsibility of the parent or the individual to purchase and consume the product advertised, is the cause—not the advertisement.

So, whatever happened to parental responsibility? I would argue that the obesity issue is initially a parental responsibility, and, later, a personal responsibility. The only role government should play is to provide the necessary information through the educational process to equip individuals with the knowledge to manage their bodies, make wiser lifestyle choices, and create more stringent requirements in physical education while in school. The physical skills and cognitive knowledge learned at an early age will play a major role in an individual's ability to maintain a reasonable weight later in life.

When I was in high school, the school district had a mandatory physical education program whereby, at the end of each semester, you were tested in a battery of six physical skill sets. These tests included the mile run, the shuttle run (which was an essentially a sprint to a stationary cone and back several times), sit-ups, rope climb, pushups, and the pegboard. At the end of the testing period, your scores at each station were added together to reflect your overall level of fitness, which could range from barely fit enough to survive on the planet and continue to consume food, to the exceptionally fit: those who were capable of kicking your you-know-what and taking away your food.

Now, each level of fitness was represented by a specific color of gym trunks, which you had to earn the right to wear, somewhat like a pecking order of physical fitness. You began your freshman year with a pair of gray trunks, which was considered the lowest level of fitness. Grays simply indicated you were still alive and functioning in some physical capacity, still taking in oxygen. So, it should be obvious that the goal for most students was to at least improve your level of fitness to the point where you were able to shed the dreaded grays. This required effort, but an effort which would not have been made by most students if the standards and expectations were not in place to achieve a higher level of physical fitness. This is what has been missing in public schools for decades: no expectations, no standards, and you reap what you sow. All of these color-coded physical fitness abilities are now likely considered to be politically incorrect due to the issues of how it would purportedly affect someone's self-esteem.

In 2013, the American Medical Association (AMA) declared obesity a disease, which only promotes the growing naïve mindset that the individual has little control of this issue and that the federal government, or someone other than ourselves, should take some responsibility for it.

In the article "Obesity Is Not a Disease," published in the *National Review* online by Michael Tanner of the Cato Institute July 3, 2013, he made the following relevant points:[11]

- "The AMA's move is a symptom of a disease that is seriously troubling our society: the abdication of personal responsibility and an invitation to government meddling."[12]
- "While obesity is a real problem, the AMA's move is actually a way for its members to receive more federal dollars by getting obesity treatments covered under government health plans. And there will almost certainly be pressure to mandate coverage for these things by private insurance carriers, under both state laws and the Affordable Care Act,"[13] which, of course, has already turned out not to be too affordable, as well as essentially dead.
- "The AMA decision shifts responsibility for weight loss from the individual to society at large, while expanded Medicare and insurance coverage socialize the cost of treating obesity, thereby inviting all manner of government mischief. After all, if being fat is not our fault, the blame must lie with food companies, advertising, or other things which need to be regulated."[14]
- "Big government reduces all of us to the status of children. We have no responsibility for anything in our lives; therefore, government must take care of us. All we have to do, like children, is give up the freedom to make our own choices—good or bad."[15]

A more recent illustration of a growing international nanny state was provided in the press release "The Lancet Commission on Obesity Publishes Major New Report" from City University in London, January 28, 2019.[16] The following quote is from Professor Corinna Hawkes, Director of the Center for Food Policy at City University of London, and one of the Commissioners, regarding turning back the international epidemic of obesity:

> We need far-sighted policymakers and private sector leaders to drive forward actions that produce benefits for obesity, undernutrition, economy and sustainability.
>
> At the moment economic incentives are driving us to over-produce and over-consume, leading to obesity and climate change. At the same time many millions still do not have enough nutritious food, leading to undernutrition. It's a warped system with an outdated economic model at its core.[17]

Interpretation: it's someone else's fault. "Economic incentives are driving us to . . . over-consume, leading to obesity."

In the same report from the Commission co-chair, Professor Boyd Swinburn of the University of Auckland:

> Until now, undernutrition and obesity have been seen as polar opposites of either too few or too many calories. In reality, they are both driven by the same unhealthy, inequitable food systems, underpinned by the same political economy that is single-focused on economic growth, and ignores the negative health and equity outcomes.[18]

Interpretation: again, it is someone else's fault. Neither Commissioner mentions personal or parental responsibility or education. They feel there must be a global accountability of food manufacturers and governments to control obesity and what Dr. Swinburn stated as a "inequitable food system."

Inequitable food systems are not the issue. I live in Kern County, California, which has an adult obesity rate of 38.5%, according to the Kern County Public Health Services Department, Community Health Assessment for 2015–2017.[19] This rate is over 18% higher than the state average of 25%. This estimate is based upon the inherent errors of the body mass index (BMI) classification, yet, from simple observation, the county certainly has a significant overweight and obesity problem. However, there is certainly no "inequitable distribution of food" problem here that would change the pattern of obesity development or prevention.

A September 19, 2018 *Western Farm Press* article, "Kern Holds Its Top Spot as Nation's Leading Agricultural County," states that "for the second consecutive year Kern County, California beat out all others in gross agricultural value."[20] Additionally, our neighboring counties to the north, Tulare and Fresno, hold the No. 2 and 3 spots nationally, respectively. So, the issue with our obesity rates has zero to do with the quality of food availability, distribution, or the price of these products. Instead, it's all about education, choices, and responsibility, both parental and personal. Simple common sense.

Parents' Responsibility

Parents are mostly responsible for our obesity problem, not the food and beverage industry. McDonald's is not your child's parent, and McDonald's does not fail to equip us to say no to excessive food intake and chronic poor food choices, and yes to a sedentary lifestyle. Who allows the excessive TV exposure? Who allows the excessive sedentary activity? Who allows the excessive eating? Who controls the types of available food in the home? Who fails to become involved outdoors with their kids after school? Who allows the excessive amount of meals eaten outside of the home that are normally very high in calories? Who fails to say no? Who fails to lead by example? Who essentially fails to be head of the household? Who fails to be the parent instead of being a pal to their children?

In 2014, the National Institute of Child Health and Human Development published "Obesity Begins Early, Prevention Is Best, But Change Is Always Possible." This report stated:[21]

> Obesity, it appears, has something in common with smoking: once the pattern is established, it's difficult to change. A new study shows that children who are overweight or obese as 5-year old are more likely to be obese as adolescents. Other studies have shown that obese adolescents tend to become obese adults. Thus, it appears that, if a child is obese at age 5, chances are high that child will become an obese adult.[22]

On October 3, 2018, the *New England Journal of Medicine* Journal Watch published "Adolescent Obesity: When Does It Begin?"[23] This report stated:

> The greatest increases in body-mass index among overweight and obese adolescents occurred between the ages of 2 and 6 years of age.
>
> With children and adolescents becoming increasingly obese, determining critical time of weight accrual offers the opportunity to better determine interventions.
>
> To that end, German investigators conducted a population-based, longitudinal study in which 51,505 children were evaluated during the first year of life, then annually until age 14 years, and again between ages 15 and 18.

The study identified the following:

- Normal-weight adolescents generally had normal body-mass index (BMI) throughout childhood.
    - Most obese adolescents had normal BMI as infants, but 22% became overweight and 33% became obese by the age of 5.
    - 90% of children who were obese at age 3 were overweight or obese as adolescents.[24]

Let me illustrate the parenting and obesity issue with the following analogy since most obese individuals begin their excessive weight gain and poor eating habits early in life while still under their respective parents' supervision and care. Do you, if you are a parent, buy cigarettes for your children to smoke each day and begin this process as soon as they are able to learn to inhale, say, by the age of 3 (Image 5.3)? The obvious answer is no, of course not—and anyone reading this will simply scoff at what appears to be a foolish question. But I would argue that many parents unwittingly allow their children to indulge in habits just as unhealthy as smoking.

I would argue that the long-term health consequences to a child whose parents have allowed them to become overweight or obese while under their care can be as devastating, if not more so, as allowing that same child to begin smoking at the same age. Specifically, in the RAND Corporation report mentioned earlier, the authors state they:

> examined the comparative effects of obesity, smoking, heavy drinking, and poverty on chronic health conditions and health expenditures" and found that "obesity is the most serious problem. It is linked to a big increase in chronic health conditions and significantly higher health expenditures.[25]

So, if the health consequences of obesity are more problematic than smoking, why would most parents discipline their children if they caught them smoking but have no problem allowing them to chronically overeat, maintain a sedentary lifestyle, and constantly make poor food choices, which is really the heart of the obesity epidemic? Parents do not understand that their job is to drive out the foolishness of childhood behavior that would include many issues along the way, including gluttony, laziness, undisciplined behavior, etc.

Regardless of how much money the government throws at the obesity issue in the primary grades, attempting to persuade children to eat healthier and exercise more, children will only do what their parents provide them the opportunity to do. A good illustration of this point occurred years ago when I returned home after teaching one evening. I sat down for a moment to simply relax and momentarily turned on the TV. The program which happened to appear first was the popular *The Biggest Loser* weight loss program. I did not care for the program for a number of reasons, but that particular night featured a father and his son, who appeared to be in his early twenties, as contestants. The father was obviously quite large and standing next to him was his son, who was also very large and deconditioned. The producers of the show then showed a picture the father had apparently provided of the two of them standing on Santa Monica Beach, California, when the son was approximately 5 years old. At the time the photo was taken, roughly 15 or so years earlier than the televised program, the father appeared essentially the same. However, the son's physical appearance was quite different. The son appeared on the beach in his swim trunks and had the physical appearance of a miniature, well-developed wrestler, trim and muscular as any healthy young child could be. The

IMAGE 5.3   Is this responsible parenting?

son had apparently not seen the photo prior to the broadcast and after viewing it with the national audience for the first time, turned to his father and asked, "Dad, why did you do that to me?" As sad as that comment is, it is true. The son's obesity and severe deconditioning in his early twenties can be directly related to his father allowing the son to assume the father's lifestyle during the son's childhood years, instead of the father changing his lifestyle and providing more appropriate lifestyle opportunities and choices for his son.

Personal Responsibility

By the time most individuals who have grown up obese are mature enough to understand the negative effects excessive weight and obesity have on their health and physical abilities, it's often too late for anything more than modest changes. By this time, parents have simply dropped the ball, so to speak, and little Johnny, who is no longer little, is incapable of doing much about his weight due to the enormous lifestyle changes that would have to occur and the length of time he would have to maintain those changes, for it to be successful. Essentially, uneducated as to how their body functions, poorly equipped with physical skills due to the lack of development in this area for most of their lives, and bombarded by enormous quantities of food, many, if not most, obese individuals find it understandably very difficult to address this issue with long-term success. The odds are certainly against them, which is why, according to a February 25, 2019 report from *Business Wire*, the weight loss and diet control market in the U.S. alone is estimated to be worth $72 billion.[26] Most have virtually no long-term success, and the weight loss industry is very aware of the many myths and misconceptions the average consumer embraces, making them easy to exploit. Most consumers simply lack understanding of the basic science skills needed to sift through all the hype and pseudoscience and recycle themselves through one failed program after another.

So, what are the basic facts we need to acquire in order to guard ourselves from being exploited by constant weight loss scams, minimize weight gain as we age, and possibly produce a significant loss in weight if able to maintain a lifestyle change for a long enough period?

Before I provide the simple basics on body fat mass, let me give you just one illustration of how straightforward this issue really is. Prior to the ever-expanding electronic distractions which would entertain children inside the home sedentarily, most children would spend most of their free time outdoors doing a wide variety of outdoor games, riding their bikes, sports, games, etc. Such activities would keep them physically active for hours, as well as naturally teach them a wide variety of physical athletic skills such as running, jumping, climbing, pullups, etc. Even if minimal activities occurred outside daily, this would account for at least 100 calories expended each day, and often much more depending upon the activity and time spent doing it. Now, eliminate that lifestyle, exchange it for the sedentary indoor one, and add a minimum of 100 calories to the frame of your child. Now multiply this by 365 days (one year) and then divide by 3,500 (which is the calories contained in one pound of body fat). The number you are going to arrive at is 10, or, the caloric equivalent of 10 pounds of body fat at the end of that year accumulated by simply exchanging daily outside activities with sedentary indoor ones. If this child is, say, 7 years old, and this occurs each of the next seven years, you now have a 14-year-old with possibly an excessive accumulation of body fat simply because the parents failed to monitor activity levels. Add to this the availability of a wide variety of high-calorie foods brought into the home (not forced into the home by junk food manufacturers), and you can begin to see the dilemma. For the past 20–30 years, you have children who burn 100–200 calories per day less than prior generations and consume several hundred calories more per day due to poor parenting. So, even if schools required more stringent physical education programs, this will never overcome a high-calorie diet that is so prevalent, by choice, as well as an excessive sedentary lifestyle for most in industrialized societies, by choice. In other words, as stated by the young man on *The Biggest Loser* program to his dad: "why did you do that to me?"

Obesity is not a disease, as the American Medical Association would like you to believe. It does not sneak up on you overnight like some viral or bacterial infection. As the research illustrates, most develop it through years of parental or personal irresponsibility or a combination of both. The obvious answer is education, disciplined behavioral habits, realistic goals, and understanding the responsibility as a parent and as an individual.

Education—Body Fat Basics

Understanding the following facts about body fat cell structure and function should help you understand what you are up against and why losing weight can be so difficult. Keep in mind that most of the world's population relies on these well-designed fat cells to stay alive during periods of food shortages. In industrialized countries, however, these same fat cells become an albatross for many due to excessive food intake and limited physical activity. As I pointed out earlier, it only takes 100 calories more than you need each day from any food sources (not just chips, sodas, fries, etc.—the typically demonized products), and over a period of one year, this could theoretically accumulate to a significant amount of unwanted body fat. Body fat, as you're going to find out, is physiologically far more difficult to get rid of than accumulate. Most people have shared the experience of the joy of putting it on, and the misery of taking it off.

The Dilemma

There is some debate whether as an individual gains more fat mass, if the fat cell just increases in size and mass, or number as well (hyperplasia). According to my reference, *Essentials of Exercise Physiology 3rd edition*, published by Lippincott

Williams & Wilkins in 2006, "a biological upper limit exists for fat cell size. After reaching this size, cell number probably becomes the key factor that determines the extent of extreme obesity. Even doubling the size of normal fat cells would not account for the tremendous difference in the fat content of the obese and nonobese people. It seems reasonable to conclude, therefore, that the excessive adipose tissue mass in severe obesity occurs by fat cell hyperplasia. As a frame of reference, an average person has about 25–30 billion fat cells. For the moderately obese, this number ranges between 60 and 100 billion, whereas the fat cell number for the massively obese may increase to 360 billion or more."[27] So, not only does the mass or volume of the fat cell increase by roughly 35%, but the numbers as well.

Now, according to several sources, the caloric equivalent available in these fat cells for energy use ranges from 30,000 to 60,000 calories, but likely considerably more, which would allow for the energy storage to walk 500–1,000 miles if fat mass were the only source of fuel.[28,29]

The most prominent misunderstanding of fat mass is that it contains a significant amount of water. Most consumers believe the water content of fat mass ranges from 30–70% by weight. However, the water content of a fat cell is roughly 5–10% as part of the cytoplasm of the cell, but the rest is all fat (Image 5.4).

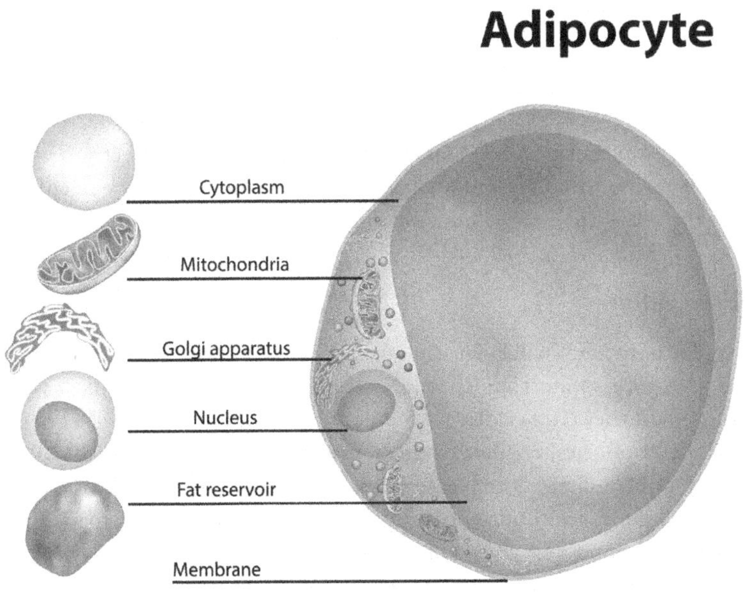

Adipocyte

Cytoplasm

Mitochondria

Golgi apparatus

Nucleus

Fat reservoir

Membrane

IMAGE 5.4 Basic composition of fat cell.

So, your effort to "sweat it off" is a waste of time. The volume of water that pours through your sweat glands during exercise or any activity is coming from your cardiovascular system and muscle tissue, not fat tissue. The muscle tissue is 70% water by weight, and that water volume is used to carry the heat you produce during exercise away from the core of the body so that you do not overheat and can maintain normal internal body temperatures. If you overheat, the brain releases hormones that in effect will make you very lethargic to prevent you from continuing to exercise and prevent the further production of heat that would damage the internal organs. Constantly replacing lost fluids helps to prevent this.

Additionally, if you fail to replace the lost fluid, the muscle tissue becomes dehydrated and your ability to actually "burn fat" will reduce dramatically with as little as a 3% loss in body weight from dehydration. Muscle tissue must be well hydrated to work at full capacity and therefore burn fat mass. So, the main point here is to prevent any significant fluid loss by frequently ingesting fluids during exercise. Simply pay attention to your thirst mechanism. Any significant loss of body weight after a training session is, for the most part, fluid loss and not fat loss, and needs to be replaced as soon as possible.

Now, Just How Much Activity Will It Take to Lose One Pound of Body Fat?

The first explanation will cover the typical adult who is sedentary, and it will be clear why we have such an obesity problem. It will also be rather depressing, because you will see what a hole you can dig for yourself, or for your children, by developing poor lifestyle habits and the difficult task ahead of you to reverse their effects.

The second explanation will highlight what a fit adult can expect and why this population tends to find itself in a relatively easy pot-hole—not the deep well so many of the obese sedentary adults and children find themselves in.

The Unfit Adult

The average unfit adult will burn 100 calories walking one mile. Most adults walk at a 3.0 mph pace, so most adults could cover three miles in one hour of walking or burn 300 calories (100 calories × 3 miles). Unfortunately, a pound of body fat contains 3,500 calories, enough to theoretically cover 35 miles (3,500/100). So that one hour of walking does not dramatically affect your total body fat content that day. Walking for an hour per day will add up and make a dramatic impact, long term (10 pounds per year), but not on a day-to-day or week-to-week basis, which often immediately discourages many obese individuals early in their effort to shed a few pounds.

So, if the average adult were to walk at a 3.0 mph pace, for one hour per day, seven days per week, this would accumulate to 21 miles (three miles per one-hour session × 7 hours per week). This would result in roughly 2,100 calories (21 × 100 calories/mile) used during exercise, not even a full pound of body fat (3,500 calories). Therefore, a reduction in food intake, roughly 500 calories per day, must be undertaken for any real hope for significant fat loss for most previously sedentary adults. The 500 reduction in calories per day would add up to 3,500 per week (1 pound of body fat), in addition to the 2,100 expended in walking resulting in a 1–2 pounds per week fat loss for most. This may not seem like much, but this is where the patience and discipline issues come into play. Over the course of a year, this could result in a significant loss in body fat mass (roughly 52 pounds).

Therefore, you can see why commercial weight loss programs which promise unrealistic results are so popular. When the realization of how hard this process is becomes a reality for the individual who is deconditioned, overweight, has no or very limited athletic skills, and possesses limited understanding as to how their body functions; they are easily victimized with unrealistic weight loss schemes. It should also be pointed out that the more obese the individual is, the more unlikely that long-term positive results will be obtained. Parents should keep this in mind when they address their children's daily energy needs.

A thin adult who is deconditioned does not normally find movement difficult or uncomfortable, just slow due to the deconditioning. So, their conditioning program progresses forward with reasonable effort. However, the more overweight an individual is, the more physically uncomfortable they are with any movement. They are not only discouraged with the deconditioning; they are discouraged from the actual effort itself—often even walking can be uncomfortable, a major deterrent to continue, which normal-weight individuals do not have to contend with. Consider being asked to spend an hour doing something seven days a week that is physically uncomfortable the entire time. How long do you think most individuals are going to be able to discipline themselves to continue? Not many. This is an issue most obese contend with, in addition to the chronic lethargy excessive weight precipitates. Excess weight essentially steals your energy levels and for many their ability to be physically productive.

The Fit but Slightly Overweight Adult

In the previous section, I illustrated the lower end of calories that can be utilized during a one-hour training/exercise session. That information was obviously discouraging for the obese population group, but it does not have to stay that way. The calories one can burn while exercising for one hour can range from the roughly 300 calories per hour to 1,200 calories an hour for someone young and well trained. This fact illustrates why those who stay conditioned have so much more latitude in the amount and types of foods they can consume without becoming overweight. It is referred to as discretionary calories. When these individuals become slightly overweight and wish to address the issue, they can do so with relative ease as compared to their sedentary, obese counterparts.

This point should also highlight the arrogance many fit adults may acquire towards the obese. Do not misunderstand: I am not in any way trying to justify poor physical stewardship, but what I am pointing out is that, likely, most obese adults were obese as children because of poor parenting. They find themselves in a pickle, so to speak, reaching adulthood with the desire to change but without the physical and cognitive skills to make any but very slow progress towards their goal. What the normal-weight adult does not understand is that as fat mass begins to shrink in size, the fat cells send out a hormone, ghrelin, to the brain to stimulate appetite to replenish what was lost. For a modestly overweight adult, this mechanism of "nagging" from the brain to eat more can be overridden for a brief few weeks to a couple of months, if necessary, with discipline. However, for the obese who are looking at years of significant lifestyle changes and reduced food intake, this "nagging" to consume more is hard to suppress for the given length of time. It is really quite an accomplishment for someone obese to trim down to a healthier weight and keep it off long term—bravo to you, for this is truly quite an accomplishment.

Physical Appearance—The Roadblock to Treating Obesity

Because of the difficulty of shedding those unwanted pounds, most overweight individuals tend to give up the effort after a very brief period due to the lack of significant visual changes in their physique. However, what must be kept in mind are the additional benefits associated with exercise and better food choices, such as increased energy levels, strength, physical productivity, control of diabetes, reduction in health care costs, etc. So, the need to focus on function vs. weight loss should be the primary reason for exercise, not necessarily weight loss, recognizing it as a responsibility one has towards themselves and others.

Some examples of improvements in quality of life absent of any significant weight loss can include:

- Improved balance.
- Improved blood sugar levels in diabetics.
- Improved blood pressure.
- Increased strength.
- Better sleep.
- Improved stamina.
- Greater productivity at work.
- Physical independence as you age.

The Downside of Rapid Weight Loss Programs

Weight loss from caloric restriction alone, of more than two pounds per week, is an unrealistic goal for most, and the physiological consequences can be negative—both short term and long-term—in either achieving your goal or maintaining weight loss. For example, a female student attending my evening nutrition course I teach at the local college recognized that her 27-year-old male bodybuilder boyfriend was experiencing the symptoms we were discussing in class regarding those who go from adequate calories to excessively low calories. She realized that her boyfriend's inability to sustain his muscle mass, as well as all the lethargy and early fatigue he was experiencing and complaining about, may be related to what we were discussing in class. He apparently had dropped his daily caloric intake to only 1,800 calories to cut weight for a competition; a considerably lower intake from the roughly 3,500–4,000 he would have needed to sustain his normal training regimen. She had him examined by a physician, and his testosterone levels had dropped to 92ng/ml, well below the normal ranges of 241–827ng/ml for males. In other words, his caloric level was so low for his needs that his brain responded as if he were anorexic. His testosterone levels dropped to prevent any further muscle tissue development and significantly reduced his energy levels (lethargy). The loss of muscle mass during this time was illustrated by a molecule measured in the blood (creatine kinase) indicative of muscle tissue damage, which was two times higher in him than the upper limit of normal. Essentially, his need for sugar to sustain the energy needs of his vital organs, central nervous system, immune system, muscle tissue, and the brain were not being met through his diet, so his body resorted to breaking down muscle tissue for the fuel source it was deficient in. Muscle fibers can act as the carbon skeleton that can be converted to energy by the liver when not enough carbohydrates (glucose) is being provided by the diet.

Resting Metabolic Rate (RMR)

Long-term rapid weight loss programs can have a profound negative effect on your resting metabolic rate. The resting metabolic rate is the body's total caloric requirement over a 24-hour period to sustain all its functions at rest, including vital organ functions, maintaining body temperature, thought processes, and respiration. In 1993, researchers examined 328 healthy men (17–80 years old) and 194 women (18–81 years old) volunteers were characterized for RMR, body composition, physical activity, peak oxygen consumption (peak VO_2), anthropometrics, and energy intake. Measured RMR was 23% higher in men (1,740 +/− 194 kcal/day) than in women (1,348 +/− 125 kcal/day). These figures depend upon body size, age, muscle mass, and climate, where the rate is lower in warmer climates.[30] You can positively change your RMR by becoming more physically active and increasing the amount of muscle mass which results in the utilization or burning more calories per day at rest than normal. If this, through increased activity, is a change of just 100 calories per, then this can theoretically equates to 36,500 calories per year (100 × 365 days) or roughly 10 pounds of body fat (36,500/3,500 calories in one pound of body fat). You can negatively affect your RMR (reduce it) by a sedentary lifestyle, losing muscle mass, or excessive caloric restriction.

The reduction in your resting metabolic rate is a natural protective mechanism by the brain to prevent what it interprets as insufficient calories to maintain vital organ function and all the related physiological functions that take place 24 hours per day, seven days a week. I observed this while running the exercise physiology lab years ago for a large medical practice. Two female employees had decided to pursue an 800-calorie-per-day liquid weight loss program through a local endocrinologist due to their excess weight. Both ignored my warning as to what will likely occur to their BMR as a result, and in the long run may possibly make it very difficult to both keep the weight off as well as maintain the weight loss. Prior to the start of their quest with the 800 calories, they both accepted my invitation to have their BMR measured in the lab; both fell into the normal range of 1,200–1,300. One month into their weight loss efforts, they both accepted my offer for a reassessment. Both their BMRs declined to roughly 950–1,000 calories per day. I was unable to acquire any long-term follow-up measurements.

Many weight loss programs will use various equations to predict one's RMR. However, several studies have demonstrated these equations to be inadequate to provide accurate information, which can greatly mislead participants. One study, reported in *Medicine and Science in Sports and Exercise* in 2014, illustrated that this would likely lead to an overestimation of the RMR of approximately 10% for men and 15% for women and possibly as high as 20–30% for some demographic groups.[31] In another study reported in *Obesity Research and Clinical Practice* in 2016, researchers concluded that prediction equations for RMR to within 10% of an actual measured value was only accurate approximately 40% of the time, regardless of gender and weight classifications. The authors concluded, "in clinical weight management settings direct measures of RMR should be made wherever possible."[32] Also in 2017, in the journal *Nutrients*, researchers concluded that prediction equations overestimate resting metabolic rates and energy requirements.[33]

Fat Cell Numbers

It is a common misconception that losing body fat results in the loss of fat cell content as well as the fat cell itself. For example, an overweight adult weighing 328 pounds has roughly 75 billion fat cells; if his weight dropped to 165 pounds, he would still have 75 billion fat cells, although much of it would now be empty or greatly reduced in size. Sadly, due to the availability of the large number of fat cells from prior weight gain, it will always be easier for the return of the excessive body fat than the initial development of it. Large individuals, once they have lost body fat, must maintain an exceptional level of restraint and discipline to prevent the return of it.

Due to the negative impact low-calorie diets have on the resting metabolic rate, many who adhere to those programs for extended periods of time often find themselves being able to regain body fat on intake of as little as 1,500 calories per day if they are sedentary. This is certainly not excessive food consumption, but is reflective of the negative impact and biological abuse of low-calorie diets.

The point is that you must sustain enough calories to allow for a slow loss in fat mass to prevent the negative physical adaptations that will come with rapid weight loss programs, which, in turn, will have a long-term negative impact on your efforts.

The Genetic Link

It is true that some individuals may have a genetic predisposition to obesity. However, the National Human Genome Research Institute of the NIH points out that "all human beings are 99.9 percent identical in their genetic makeup."[34] So, we are all part of the same basic genetic pool, which means this predisposition exists among people all over the world, yet we see more obesity in the United States and other industrialized countries due to poor parenting, diminished personal discipline and responsibility as previously discussed, decreased daily physical demands, greater availability of food, and excessive intake of high-caloric foods. For most, you cannot attempt to blame genetics as the reason for obesity.

So, the bottom line is that obesity is both a parental and personal responsibility issue for most individuals. The government cannot resolve it, nor is the food and beverage industry responsible for it. We all make choices, some good and some bad. While it is true that the temptations for excess are more prevalent now than in the past, it is still a choice to indulge or deny these temptations.

Some Tips for Sensible Weight Control

The most effective and lasting way to reduce body fat is to increase physical activity and decrease calories. Here are some simple rules of thumb that will help eliminate the 300–500 calories per day necessary to shed body fat:

- Eat at home as often as possible. This will help you avoid the high-fat and calorie-rich foods on restaurant menus. It takes a lot of discipline to turn down a tasty BLT and a chocolate shake. Try not to eat away from the home more than two times per week.
- Do not keep addictive snack foods (chips, candy, ice cream, etc.) in the house unless you can keep portion sizes reasonable. Pick two or three nights a week to have one of these with a meal or as dessert.
- Try to prevent late-night grazing. Brush and floss your teeth soon after dinner. The hassle of re-brushing can be enough to prevent the 10 p.m. refrigerator raid.
- Try to be active every day. The activity doesn't have to be intense; expending even an extra 100 calories a day will account for 10 pounds a year (100 calories × 365 days = 36,500/3,500 calories = 10.4 pounds). Simply turn

off the TV and find something more constructive to do with your time, like walking a mile. Just get up and move.

- Do not assume you have to have breakfast. This is a myth. Most individuals leave their homes each day for a white-collar job of some type which requires very little energy output physically. Your muscle tissue stores more than enough sugar for its use, the liver can release its stored sugar to maintain brain function, and your fat mass is loaded with calories your body can draw from as well. The problem for many is that when they do become hungry later in the morning, about 10 a.m., they tend to snack on inappropriate foods around the office, such as doughnuts. To avoid this, simply bring something from home to consume when you feel like eating. Skipping breakfast does not apply to children heading off to school.

Things to Avoid

Sensible weight loss or weight control is achieved only with good choices— choices about not only food and exercise, but about diets and diet aids. While you're committing to physical activity and better eating habits, be sure to avoid the following:

- Weight loss pills or products. The FDA routinely identifies tainted weight loss products which contain pharmaceutical grade drugs or unlabeled stimulants. According to the 2019 FDA review "Tainted Weight Loss Products," the agency has "identified an emerging trend where over-the-counter products, frequently represented as dietary supplements, contain hidden active ingredients that could be harmful. Consumers may unknowingly take products laced with varying quantities of unapproved prescription drug ingredients, controlled substances, and untested and unstudied pharmaceutically active ingredients. These deceptive products can harm you! Hidden ingredients are increasingly becoming a problem in products promoted for weight loss."[35] The FDA also warned of the following: "Remember, FDA cannot test all products on the market that contain potentially harmful hidden ingredients. Enforcement actions and consumer advisories for tainted products only cover a small fraction of the tainted over-the-counter products on the market."
- High-protein or high-fat diets (Paleolithic or caveman diets). These regimens deplete stored sugars in the muscle, which inhibits the muscles maximal ability to utilize or burn the stored body fat. The optimal method for the removal of stored body fat is to allow the muscle tissue the capacity to work at high intensity levels for extended periods of time. In the absence of stored sugar due to high-protein/low-carbohydrate diets, the muscle tissue is unable to work at full capacity or for extended periods of time. Both

prevent the optimal loss of body fat, long term. What's more, the diuretic effect that often accompanies high-protein diets inhibits the ability of muscle tissues to dissipate heat generated during exercise; to prevent overheating, the brain through hormonal activity will diminish muscle activity, which causes lethargy and prevents the optimal utilization of body fat, as well. Staying well hydrated, as well as maintaining enough carbohydrates in the diet, is essential for losing and maintaining fat loss long-term. Diets that eliminate or limit complex carbohydrates eliminate thousands of plant chemicals found in carbohydrate foods that are essential for good health as well. As an example, certain healthy bacteria in the colon rely on carbohydrates to proliferate and maintain the health of the colon. Without enough healthy carbohydrates, your risk for colon cancer will increase (see ketogenic diet information in Chapter 10).

- Diets that fall below 1,200 calories for sedentary females or 1,500 calories for sedentary males. Individuals who repeatedly do this will find that they will eventually gain weight if they eat more than 1,200 calories per day.
- Meal replacement wannabes. These typically are unbalanced and costly and have no long-term benefit. The so-called "meal replacement" items are especially misleading. For example, a meal replacement product might include beta-carotene on its ingredient label, but food includes hundreds of other important carotenoids that the product does not contain. These products cannot take the place of food, and claims that they do are simply deceptive hype.
- Energy drinks. From personal experience from taking literally hundreds, possibly thousands, of blood pressure readings over the years, it is not uncommon for apparently healthy young males who we conduct pre-employment screens on, for physically demanding jobs in the oil fields and elsewhere, to have resting blood pressures above 160/100 as a reaction to the consumption of energy drinks. This self-induced high blood pressure is strongly associated with memory problems later in life, as well as possibly dementia and Alzheimer's, in addition to the cardiovascular risks associated with these products. In 2012, a study was published in *The Lancet Neurology* which detected changes in brain anatomy, white-matter integrity, of those with even mildly elevated blood pressures.[36]
- Sauna suits. Body fat contains relatively little water, so the fluids lost with the use of sauna suits come from muscle and blood volume. The resulting dehydration will only hinder fat loss efforts by reducing the optimal functioning of your cardiovascular and musculoskeletal systems, which in turns reduces the overall oxidation (burning) of body fat, as stated earlier. Staying well hydrated is essential for optimal body fat loss.
- Creams advertised to remove "cellulite" (just another marketing name for "fat"). Cellulite is the dimpled fat on the buttocks or thighs of women. This fat tissue is no different than fat tissue found elsewhere on the body. The

dimpling or bulging appearance of this fat tissue is attributed to the bands of fibrous tissue located in this area.

- Anything promoted with terms like *miraculous, fast, easy, effortless, while you sleep,* and so on. Most people will agree that it certainly is "miraculous, fast, easy, effortless, it happened while I slept, etc." to put on excess body fat, but the opposite is not true, and it never will be. It is far easier to prevent excessive accumulation of body fat than it is to take it off. If most weight loss programs were even remotely successful, the current international problem with obesity would be minimal.

Notes

1. C.M. Hales, M.D. Carroll., C.D. Fryar, and C.L. Ogden, Center for disease control and prevention report. *National Center for Health Statistics* (October 2017), No. 288.
2. J. Cawley and C. Meyerhoefer, The medical care costs of obesity: An instrumental variables approach. *Journal of Health Economics* (2012), Vol. 31, No. 1, pp. 219–30.
3. https://www.cdc.gov/healthyweight/effects/index.html
4. R. Sturm and K.B. Wells, *The health risks of obesity: worse than smoking, drinking or poverty*. Santa Monica, CA: RAND Corporation (2002). https://www.rand.org/pubs/research_briefs/RB4549.html
5. C.A. Raji, et al., Brain structure and obesity. *Human Brain Mapping Online Version* (March 2010), Vol. 31, No. 3, pp. 353–64.
6. J. Brody, Attacking the obesity epidemic, first figuring out its cause, *New York Times* (September 12, 2011).
7. Ibid, p. 1.
8. Ibid, p. 1.
9. L. Donnelly, Pizza must shrink or lose their toppings under government anti-obesity plan. *The Daily Telegraph* (October 11, 2018).
10. Ibid, p. 2.
11. M. Tanner, Obesity is not a disease. *National Review* (July 3, 2013).
12. Ibid, p. 1.
13. Ibid, p. 2.
14. Ibid, p. 2.
15. Ibid, p. 4.
16. S. Cubbin, The Lancet Commission on Obesity Publishes Major New Report. *City University in London* (January 28, 2019).
17. Ibid, p. 1.
18. Ibid, p. 2.
19. http://kernpublichealth.com/wp-content/uploads/2017/04/Community-Health-Assessment-2015-2017.pdf
20. Kern holds its top spot as nation's leading agricultural county. *Western Farm Press* (September 19, 2018).
21. Obesity Begins Early, Prevention Is Best, But Change Is Always Possible. *National Institute of Child Health and Human Development* (2014).
22. Ibid, p. 1.
23. Geserick M. et al., Acceleration of BMI in early childhood and risk of sustained obesity. *New England Journal of Medicine* (October 4, 2018), Vol. 379, p.1303. https://doi.org/10.1056/NEJMoa1803527
24. Ibid, p. 1.
25. R. Sturm and K.B. Wells, *The health risks of obesity: worse than smoking, drinking or poverty*. Santa Monica, CA: RAND Corporation (2002) p. 1. https://www.rand.org/pubs/research_briefs/RB4549.html

26. The $72 billion weight loss & diet control market in the United States, 2019-2023 – why meal replacements are still booming, but not OTC diet pills-researchandmarkets.com. *Business Wire* (February 25, 2019). https://www.businesswire.com/news/home/20190225005455/en/72-Billion-Weight-Loss-Diet-Control-Market

27. W.D. McArdle, F.I. Katch, and V.L. Katch, *Essentials of Exercise Physiology,* 3rd ed. (Lippincott Williams & Wilkins in 2006), p. 595.

28. T. Purdom, L. Kravitz, K. Dokladny and C. Mermier, Understanding the factors that effect maximal fat oxidation. *Journal of International Society of Sports Nutrition* (January 12, 2018), Vol. 15, No. 3, p. 3.

29. E. Coyle, Fat metabolism during exercise: New concepts. *Sports Science Exchange* #59 (1995), Vol. 8, No. 6, p. 2.

30. P.J. Arciero, M.I. Goran, and E.T. Poehlman, Resting metabolic rate is lower in women than in men. *Journal of Applied Physiology* (December 1, 1993), Vol. 75, No. 6, pp. 2514–20.

31. R.G. McMurray, J. Soares, C.J. Caspersen, and T. McCurdy, Examining variations of resting metabolic rate of adults: A public health perspective. *Medicine and Science in Sports and Exercise* (July 2014), Vol. 46, No. 7, pp. 1352–58.

32. T.G. Wright, B. Dawson, G. Jalleh, and K.J. Guelfi, Accuracy of resting metabolic rate prediction in overweight and obese Australian adults. *Obesity Research and Clinical Practice* (September 2016), Vol. 10, Supplement, pp. S74S83.

33. R.T. McLay-Cooke, A.R. Gray, L.M. Jones, R.W. Taylor, P.M.L. Skidmore, R.C. Brown, Prediction equations overestimate the energy requirements more for obesity-susceptible individuals. *Nutrients* (September 13, 2017), Vol. 9, No.9:1012. doi:10.3390/nu9091012

34. Genetics vs. Genomics Fact Sheet. *NIH National Human Genome Research Institute* (September 7, 2018), p. 2. https://www.genome.gov/about-genomics/fact-sheets/Genetics-vs-Genomics

35. Tainted Weight Loss Products. *Food and Drug Administration* (September 10, 2019), p. 1.

36. P. Maillard, S. Seshadri, A. Beiser, J.J. Himali, R. Au, E. Fletcher, et al., Effects of systolic blood pressure on white-matter integrity in young adults. *The Lancet Neurology* (December 1, 2012), Vol. 11, No. 12, pp. 1039–47.

6

PROTEIN NEEDS OF ATHLETES

Seven Misconceptions

There is no question that the established dietary guidelines for protein set by what are commonly known as the RDAs at 0.8 g/kg/day, are far too low for anyone whose lifestyle requires their body to function beyond its resting metabolic levels. Current evidence indicates that most lifestyles beyond comatose require from 1.2 to 1.6 g/kg/day of high-quality protein per day for optimal health outcomes as well as athletic performance and physical development, and possibly 2.0 g/kg/day for some heavily training athletes. This equates roughly to 20–30 grams of high-quality protein four times per day, and possibly another consumed prior to sleep for some. For an excellent in-depth review of this, the following two reviews published in 2016, in *NRC Research Press*, "Protein Requirements Beyond the RDA: Implications for Optimizing Health," and 2018 in *Nutrients Review*, "Recent Perspectives Regarding the Role of Dietary Protein for the Promotion of Muscle Hypertrophy With Resistance Exercise Training," are recommended. However, this chapter focuses on two main points:

1. Does this higher intake level require supplemental intervention to be achieved? The answer to this is no, which will be presented, and the recommended levels of intake can be easily achieved using high-quality animal protein sources such as milk, eggs, poultry, meat, etc.
2. What misconceptions do most athletes and active consumers embrace which may lead them to believe they need protein supplemental intervention? There are many.

Regarding the first question, consumers and athletes need to keep in mind the points made in Chapter 9 regarding the potential hazards of supplements when considering their sources for protein. Also, Consumerlab.com tests revealed that

20% of the protein supplements they have tested are mislabeled. With those two points in mind consider the cost as well. Consumerlab.com states that to obtain 20 grams of protein from a supplement will range from 80 cents to $1.50. According to the December 24, 2018 USDA Agricultural Marketing Services Dairy Report,[1] the average cost of a gallon of milk in the United States is $3.27 for whole milk and $3.21 for 2% milk, both containing 128 grams of protein. So, you can purchase 128 grams of protein from milk for $3.21, without the fear of it being mislabeled or spiked with an unwanted stimulant or steroid, or, you can purchase the equivalent of that 128 grams of protein from the supplement industry from anywhere between $5.12 and $9.60.

Most athletes embrace the following seven misconceptions regarding muscle mass and their perceived protein needs, which should be addressed by those who supervise them.

1. The composition of their muscle tissue. Specifically, just how much protein is in one pound of muscle?
2. Excessive protein supplementation to insure muscle development.
3. Branched-chain amino acids. Helpful, hype or possibly harmful?
4. Is there a magical, or "proprietary," blend of amino acids, which will assist muscle development?
5. Why do many athletes feel and perform better with the use of protein supplements?
6. How much muscle development can the athlete expect in one week?
7. Can a normal diet provide enough protein?

Misconception 1: Muscle Composition—How Much Protein Is in One Pound of Muscle?

Most athletes assume muscle tissue is predominantly protein. This is false. Muscle tissue is roughly 75% water, 22% protein, and the balance glycogen (sugars), lipids (fat), and mineral salts. An easy illustration of this is the production of beef jerky. The water content of beef muscle according to the USDA, depending upon the cut, is comparable to human muscle tissue. So, simply fully dehydrate one pound of beef, and when the initial one pound of beef (muscle) is fully dehydrated, what remains will be only 20–22% of the initial muscles mass, or roughly 100 grams of amino acids, which is far less than most athletes assume. Keeping in mind that the muscle developmental process is slow, roughly 0.6 pounds per week under the best conditions, this should clearly illustrate to the athlete that their daily protein needs are not as significant as most believe. This is the reasoning why sport governing bodies, sports medicine organizations, the *Joint Position Statement of the American College of Sports Medicine and The Academy of Dietetics and Nutrition*, US Armed Forces Consensus Statement, as well as the US Military Special Operations, all recommend the well-established 20–30 grams of quality protein (those which contain all nine essential amino acids) after training sessions to meet the athlete's post-training needs, if the athlete is consuming sufficient total calories. If the athlete is not consuming enough calories, his/her protein needs will increase, due to substantially more of the ingested protein carbon skeleton needing to be used as an energy source instead of for muscle protein synthesis. This is a common problem with many athletes, as will be illustrated below.

Misconception 2: Excessive Protein Supplementation to Ensure Maximum Muscle Development

Any excess dietary protein consumed during any meal is readily broken down by the liver; it is not a health risk for a healthy adult or teenager. The nitrogen part of the protein molecule is readily excreted in the urine as urea. The remaining carbon skeleton part of the protein molecule is readily converted into fat and stored, or used as an energy source if needed, which is one of the main underlining factors why many athletes experience positive results using protein supplements. If their training regimens are not sufficiently supported with the appropriate energy intake (calories), then the "protein" supplement is simply filling the energy void by providing the missing calories. This, of course, enhances the athlete's training, as well as development, which the athlete misinterprets as coming from the protein directly. This is a classic misunderstanding. The improvement in performance or development is related to the extra energy (calories) the product (protein and related carbohydrate ingredients) may provide, which would be better obtained from carbohydrate food sources, such and produce and grains. Misconception 5 will explain this further.

However, excessive protein through supplementation may hinder muscle development and recovery in two ways:

Point 1. Exacerbating dehydration which may occur with training, unless enough fluid is ingested to compensate. Specifically, when the excess protein consumed is broken down, as explained previously, the excess nitrogen, which is part of the protein molecule, must be excreted. To do so, the liver transforms the excess nitrogen into urea and sends it to the kidneys for excretion in the urine. However, the kidneys must dilute the higher concentration of urea now found in the urine, resulting in an increased fluid loss, which as Peter Lemon, Ph.D., has pointed out, could result in as much as a "four to fivefold increase in urine volume"[2] in some could produce a diuretic effect. So, attention to the appropriate fluid intake for athletes consuming a higher protein intake is essential. Remember, muscle tissue is mostly water by weight, so any dehydration of muscle will significantly affect muscle tissue performance, as well as recovery, possibly inhibiting the intensity of successive training days. Additionally, researchers at the University of Connecticut's Department of Nutritional Sciences who studied hydration levels in athletes, presented their findings at the April 2002 meeting of the Federation of American Societies for Experimental Biology. The researchers provided athletes with either 68 grams, 123 grams, or 246 grams of protein daily for four weeks. Those athletes consuming the greatest amounts of protein demonstrated significantly lower hydration levels.[3] Just a 2% dehydration level may negatively impact physical performance.

Point 2. To understand the second point, the reader must understand a point made in Chapter 8 regarding the supplemental ingestion of antioxidants for the purpose of suppressing free radical production after exercise. The old assumption was that the free radicals produced during exercise were a negative consequence of exercise, and therefore should be inhibited. However, it is now well recognized that those free radicals produced during exercise are a signaling mechanism which precipitates muscle adaptation to the stress load. So, inhibiting this mechanism is obviously ill advised. Ironically, many athletes consume protein supplements in considerable excess, through supplementation, which may have the same negative effects, but through a different mechanism of action, as pointed out in the 2018 protein metabolism review by Stuart Phillips, Ph.D., et. al., Department of Kinesiology, McMaster University, Hamilton, Ontario, in the *Nutrients Review*. Here are three specific relevant points taken from this review:[4]

- "Increased protein turnover (muscle tissue breakdown) appears to be necessary, especially during the early resistance-training period, to facilitate skeletal muscle remodeling and to lay the foundation for subsequent muscle protein accretion (the process of growth or increase) with progressive training. The impact of suppressing this normal rise in muscle protein breakdown is not known, however, it appears unlikely, in our estimation to translate into any physiological benefit" (page 6).

- "Consumption of greater quantities of protein per meal than what we are recommending here (i.e., ~20–30 g/meal) may suppress proteolysis (the breakdown of proteins or peptides into amino acids by the action of enzymes). We see little evidence to support strategies that aim to specifically suppress muscle protein breakdown following resistance exercise due to the role muscle protein breakdown would play in protein remodeling during recovery from exercise and recovery and because of our relative lack of understanding of the potential consequences of doing so" (page 7).

- "*In our view athletes and recreationally active individuals should focus on practices to maximize muscle protein synthesis rather than suppress muscle protein breakdown, which appears to be physiologically important for skeletal muscle remodeling following damaging exercise*" [my emphasis] (page 8).

The take-home message: no benefit has been observed with higher levels of protein intake, but the possible negative consequences may be the hindrance of muscle tissue adaptation, just as with supplemental antioxidants (see Chapter 8).

Misconception 3: Branched-Chain Amino Acids—Helpful, or Hype and Possibly Harm?

There is a total of 20 amino acids which are required for the structure of any muscle protein. Eleven of these are considered non-essential because your body synthesizes them. The remaining nine must be present during the process of muscle synthesis or development. The absence of any one of these nine will prevent this synthesis from taking place. The branched-chain amino acids (BCAAs) are only three of the nine essential ones needed for synthesis of new muscle tissue. So, it is completely illogical to assume that you can manufacture new muscle fiber without all the materials (all nine essential amino acids, not just three) to do so. An analogy would be to approach an architect to build a new home using only 33% of the needed material. See what kind of a response you receive. It will not be flattering.

Now, the hype for the very popular BCAAs to increase the synthesis of new muscle tissue is not only physiologically illogical, but unfounded experimentally once you eliminate the junk science, or the misinterpretation of a very transient signaling effect of their ingestion within the cell, which does not result in any long-term net gains in muscle mass. Here are some points to consider before purchasing BCAAs as part of your training regimen. The first point is taken from an excellent 2017 review of this topic in the *Journal International Society of Sports Nutrition*, titled "Branched-Chain Amino Acids and Protein Synthesis in Humans: Myth or Reality?"[5] The second point is taken from a 2016 report in the journal *Nature Medicine*, titled "A Branched-Chain Amino Acid Metabolite Drives Vascular Fatty Acid Transport and Causes Insulin Resistance."[6]

1. "When all evidence and theory is considered together, it is reasonable to conclude that there is no credible evidence that ingestion of a dietary supplement of BCAAs alone results in a physiologically-significant stimulation of muscle protein. In fact, available evidence indicates that BCAAs decrease muscle protein synthesis. All essential amino acids (9) must be available in abundance for increased anabolic signaling to translate to accelerated muscle protein synthesis."[7]

2. "Epidemiological and experimental data implicate branched-chain amino acids (BCAAs) in the development of insulin resistance, but the mechanisms that underlie this link remain unclear. Insulin resistance in skeletal muscle stems from the excess accumulation of fat," a process they feel they have now identified, and it is related to one of the BCAA, valine. The increased insulin resistance may involve a metabolite of valine called 3-HIB. This compound

"stimulates muscle fatty acid uptake and promotes lipid accumulation in muscle, leading to insulin resistance." Researchers feel their research provides an explanation for how increased BCAA intake may cause diabetes. This is preliminary, of course, and needs to be replicated.[8]

The take-home point here is simple. BCAAs will not enhance your development, and their excessive intake through supplementation may increase your risk for hyperglycemia and type 2 diabetes if further research replicates the findings. Recall *the Principle of Toxicology—the dose makes the poison*, discussed throughout this book. Any chemical compound, including all those your body requires to function normally, have an upper dose of safety that when surpassed may become harmful vs. helpful.

Misconception 4: Is There a Magical or "Proprietary" Blend of Amino Acids That Will Assist Muscle Development?

This is a total con. The first time I heard this phrase was during an NBC *Dateline* piece I was asked to appear on in 1996 called "Hype in a Bottle." During the airing of the show, the CEO of a supplement company at the time stated, "it contains a special blend of proprietary amino acids which assists in muscle development," yet he could produce no clinical studies to illustrate that his special concoction was special at all, just hype.

There are 20 amino acids which are needed to make a complete protein molecule, as stated previously. Eleven of these your body makes, and nine must be supplied by the diet and are present in all animal products, as well as grains and vegetables, in various amounts. No supplement company can produce any combination of amino acids needed by your muscle tissue which are not already present in food. Proprietary blends of amino acids are just another way of exploiting the misinformed athlete. Simply drink more milk, have some eggs, indulge in a chicken or beef sandwich, or enjoy some beans and rice.

Misconception 5: Why Do Many Athletes Feel and Perform Better With the Use of Protein Supplements?

Many athletes will argue that they obtain a significant benefit from their protein supplement, which is true. But is this benefit related to some special "proprietary" blend of amino acids, the protein itself, or is it more directly attributed to some other variable? For most consumers who utilize protein supplements, the perceived benefits of them are far more likely to be directly related to one, or a combination of, the five following variables, and the supplement industry is very good at exploiting these misunderstandings the typical consumer overlooks.

The Four Variables Why Protein Supplements Appear to Work:

1. A carbohydrate-deficient diet
2. Natural progression
3. The placebo effect
4. Stimulants and steroids

Variable 1—A Carbohydrate-Deficient Diet

Carbohydrates are the preferred energy source for muscle, the central nervous system, brain, and immune system, to name a few of the bodily functions which rely on them as their main energy source. So, it should be obvious that if this primary energy source is unavailable, which is common for many athletes and active consumers, there is going to be some negative consequences physiologically and developmentally, as well as with physical performance. Here is a list of four consequences most will notice:

1. Inability to sustain intensity levels or duration due to the lack of stored sugars.
2. Inability to gain muscle mass due to a reduction in testosterone levels.
3. Decreased immunity.
4. Decreased coordination, reaction time, concentration, and increased incidence of injury.

Let me clarify each of these points.

1. Inability to Sustain Intensity Levels or Duration

This is best illustrated using Image 6.1.

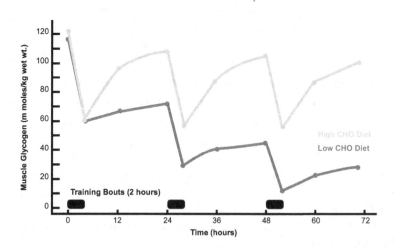

IMAGE 6.1 Effects of a low–carbohydrate diet on glycogen stores as training week progresses.

In the graph, the first downward sloping lines represent the amount of stored sugar utilized during the first practice of the week on Monday. Between Monday and Tuesday, there are two upward sloping lines, one marked High CHO Diet and the other Low CHO Diet. The low carbohydrate line represents the typical diet of many athletes, with limited carbohydrate intake over the 24 hours following Monday's practice. The high carbohydrate line represents the amount of sugar replenishing that would have taken place if the carbohydrates in the diet had been sufficient. Clearly, those athletes consuming a higher carbohydrate diet significantly replenish their levels of muscle fuel (sugar) between practices, while those on a lower carbohydrate diet do not.

On Tuesday, or the 24th hour mark on the graph, both downward sloping lines again represent the amount of stored sugar, or glycogen, that will be used during the next practice session (day 2). Those on a lower carbohydrate diet start each successive practice with significantly less available muscle fuel (sugar), compared to those in the higher carbohydrate group.

By following the progression of the lines through Thursday and Friday, it is easy to see that those athletes who do not consume sufficient carbohydrates will have insufficient energy stores at the beginning of practice sessions or competition later in the week. Those athletes begin to demonstrate early signs of fatigue, diminished intensity, and decreased performance by Thursday—an effect that is directly related to insufficient carbohydrate intake, not protein, and not necessarily to the physical training exertion.

So, in a practical sense, what does this mean to the athlete? Athletes who consume enough carbohydrates will have significantly greater amounts of stored sugar to draw upon during the course of a training week, enabling them to sustain intensity levels and endurance, and experience greater physical development rates. This is an issue the supplement industry is very well aware of, and readily exploits. Here's how.

As the graph illustrates, the diet too low in carbohydrates gradually diminishes the available stored sugar for muscle tissue to use as the week progresses. So, at any time during the middle or end of the training week, the athlete consumes a "protein" drink, which is simply converted to an energy source, and not used as a protein, resulting in the athlete experiencing an improvement in energy level and performance. The obvious tendency then of the misinformed athlete will be to credit the improvement to the protein in the drink or supplement. But this is not the case. The benefit is simply coming from the extra calories (energy) provided by the supplement, not the protein. Any combination of carbohydrate food items, from produce or grains, would provide the same benefit if the athlete has already met their protein needs. The athlete is simply underfed calorically, which is common, and once those calories which are lacking to support the energy demands of the training are provided, regardless of its source, the athlete is going to feel and perform better, and this is exactly what the supplement industry relies upon as one of the tools to deceptively market their products.

The supplement industry understands that all it must do is to convince any athlete who consumes too few carbohydrates, to try their product, and due to the athletes misunderstanding as to what is occurring physiologically, will then experience positive training and or performance outcomes, which the athlete will wrongfully attribute to the purported benefits of the protein supplement, when it was simply the need for more calories or carbohydrates, not protein. As an example, during lectures on this topic, I use a product whose front label states the following:

- "high-quality protein"
- "branched-chain amino acids"
- "glutamine peptide whey protein"

Clearly, based upon the front label, this product is promoted as a protein supplement. However, when you look on the ingredient list on the back of the container in small print, something most younger athletes would never read, you will find that 80% of the 290 calories per serving comes from carbohydrates, with only 10 grams of protein. The protein content comparable to consuming 1 cup of milk. Is the athlete going to feel better after consuming this product? Of course. What myth is it going to perpetuate? The need for more protein, when in fact it is the carbohydrate content of this product which should be credited with enhancing the athlete's performance, and not the protein.

A recent example of this was demonstrated by Saint Louis University researchers when comparing the typical Western diet with the use of a higher carbohydrate Mediterranean diet. The study was published in the *Journal of the American College of Nutrition* on February 13, 2019, "Short-Term Mediterranean Diet Improves Endurance Exercise Performance."[9] It was only a small study using 11 recreational athletes, but it is reflective of what one would expect based upon what I have just explained. While the runners were on the higher carbohydrate Mediterranean diet for just four days, they improved their 5-km run time by 6% with similar rates of perceived exertion as well as heart rates as when they were ingesting the Western diet.

2. Inability to Gain Muscle Mass Due to a Reduction in Testosterone Levels

Most everyone is familiar with anorexia and the negative effects it has on the body, especially the loss of muscle tissue. This condition has some similarities to athletes who limit their energy intake. An anorexic will consume less than 1,000 calories per day, well below the roughly 2,000 or more calories per day for the average adult. This is a calorie per day deficit of 1,000 or more. Now consider a motivated athlete in training who may have a minimum need of 3,000 calories per day, often times much higher, but is only consuming 2,000. This, as in anorexia, is a 1,000 calorie per day deficit, and is interpreted by the brain much the same way: a false athletic anorexia. The brain regards this excessively low caloric intake insufficient to meet the energy needs of the body, which sets off a series of hormonal adaptations by the brain to try to adapt to the insufficient calories.

One of those adaptations is the significant reduction in testosterone in males and estrogen in females. This is why many males, as well as females, who consume insufficient calories, especially from carbohydrates, will chronically complain of fatigue, lack of endurance, muscle wasting or stagnation in development, etc. The reduction in testosterone makes it impossible to put on new muscle mass, and many athletes are lucky to just maintain muscles mass on the low-calorie diet. The athlete simply has reduced their eight-cylinder engine to a four-cylinder.

Let me provide an example using the 27-year-old bodybuilder mentioned in the last chapter. He was complaining of chronic fatigue, loss of strength, and size of muscle mass. He stated he was preparing for an upcoming bodybuilding show and was trying to reduce as much body fat as possible prior to the competition. In the process, he stated he consumed roughly 4,000 calories per day during normal training periods, but had reduced that caloric level to roughly 1,800–2,000 calories per day to rapidly reduce fat mass. To me, it was clear what was taking place, so I made a simple suggestion—"See your M.D. and have some lab work performed, and make sure he measures your testosterone levels," which he did.

A normal adult male would have normal testosterone levels anywhere from 241–827ng/ml, which, based upon his size and level of physical development, his would have been. However, his lab work results showed 92.6ng/ml during his excessive low caloric period. Now, it does not take a genius to figure this out. It was quite clear that due to his high caloric output level from training compared to his excessively low caloric intake, his body was responding as if he were anorexic, and therefore the major drop in testosterone levels. This is a common problem with athletes, especially at the high school and college levels. The athlete corrected his caloric levels, and roughly four weeks had passed prior to seeing his M.D. again. Upon follow-up lab work, his testosterone levels had risen to 300ng/ml, and he was obviously feeling considerably better, as well as stronger.

To support this, his M.D. also measured his levels of creatine kinase, which is an enzyme in your blood which when high can be indicative of skeletal muscle injury, as well as a heart attack and other chronic diseases. Normal levels in the blood are 22–198 U/L. His, while consuming the excessive low calories, were 420 U/L. His levels were ruled as reflective of significant muscle tissue damage (wasting) and unrelated to any other chronic medical condition or a heart attack. Clearly, due to his level of training being improperly supported by his caloric intake, his body was breaking down muscle tissue to use its carbon skeleton as an energy source as well as other needs.

Now, using this lesson as an illustration, magnify this scenario by thousands of athletes nationwide making the same mistake regarding excessively low caloric levels to sustain their physical training regimens, which the supplement industry is very aware of. When any one of these athletes happens to pick up any over-the-counter supplement, regardless if its caloric source, and consumes it, they are going to immediately experience benefits, not because of some special ingredient or "proprietary" mix of amino acids, but simply from the extra calories in the diet, which could have just as easily been met by most any other food items.

Correcting the caloric deficit, which will significantly increase their testosterone levels back to normal ranges, provides the athlete with a boost in training levels they have not experienced for some time, and they will obviously associate that improvement to the supplement vs. just the increased calories from it.

3. Decreased Immunity

The white blood cells need carbohydrates in order to proliferate normally throughout the body and aid the immune system in combating invading bacteria, viruses, and fungi. Many athletes become ill soon after increasing the intensity or duration of their training program partly due to a less responsive immune system as a result of impeding the effectiveness of their immune cells. One of the more prominent researchers in this area is David Nieman, Ph.D., of Appalachian State University, Department of Health, Leisure and Exercise Science. In the *Journal of Sports Sciences* in 2004, he and his colleagues published the review article "Exercise, Nutrition and Immune Function."[10] They state the following:

> To maintain immune function, athletes should eat a well-balanced diet sufficient to meet their energy requirements. An athlete exercising in a carbohydrate-depleted state experiences larger increases in circulating stress hormones and a greater perturbation of several immune function indices. Conversely, consuming 30–60 g (120–240 calories) carbohydrate per hour during sustained intensive exercise attenuates rises in stress hormones such as cortisol and appears to limit the degree of exercise-induced immune depression. Convincing evidence that so-called 'immune-boosting' supplements, including high doses of antioxidant vitamins, glutamine, zinc, probiotics and *Echinacea*, prevent exercise-induced immune impairment is currently lacking.[11]

In 2000, the *European Journal of Applied Physiology*, in "Training and Natural Immunity Effects of Diets Rich in Fat or Carbohydrates," stated that when ten untrained men ingesting a carbohydrate-rich diet (65%) while endurance training was performed 3–4 times a week for seven weeks, their natural killer cells activity increased on the carbohydrate rich diet but decreased on the fat rich diet.[12] The excessively fat rich diet and low-carbohydrate-rich diet suppressing immunity was also pointed out by Dr. Nieman and his colleagues in the *Journal of Sports Science* article mentioned previously. They stated, "Immune system depression has also been associated with an excess intake of fat."[13]

4. Decreased Coordination, Reaction Time, Concentration, and Increased Incidence of Injury

It is common knowledge among sports scientists that the combination of central (brain) fatigue and peripheral (muscle) fatigue brought on by prolonged exercise and insufficient glucose to maintain brain and nerve tissue energy needs significantly affects the athlete's ability to concentrate and react to external stimuli and the muscular coordination to do so. Remember, the primary fuel source for both the brain and nervous tissue is glucose (carbohydrates from produce and grains). If glucose from carbohydrates is insufficient, then diminished function of both tissues will result. All these factors increase the susceptibility to injuries, especially toward the end of a training week or a lengthy competitive event. The risk of injury is even greater in contact sports, such as football, rugby, soccer, and basketball, in which the athlete's reaction time to those around him, changes in direction, stabilization when landing after jumping, etc., play such a critical role in injury prevention.

Variable 2—Natural Progression

It's a simple concept: consistent, well-planned training produces changes in skills and strength levels. When athletes take an inert product around the same time, they would see natural progression in physical development due to training, or normal surges in muscle mass and strength because of natural increases in testosterone levels, which occur routinely through their developmental years, the athlete tends to credit the product instead of the training or natural progression (Image 6.2).

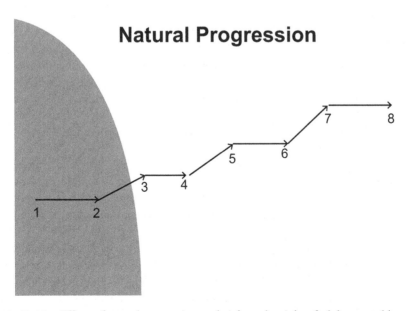

IMAGE 6.2   Effects of natural progression on height and weight of adolescent athletes.

As an example, look at the graph in Image 6.2 and follow the arrows beginning on the left. Let's say at arrow 1 to 2 the athlete may be 5-foot-7 and weigh 160 pounds at age 16. Now due to his age, this athlete may experience a natural surge in development of an inch or so in height, as well as five or so pounds of body mass over the next several months due to normal development, as indicated by the second arrow projecting upward from 2 to 3 representing this. Now if this athlete happens to take a protein supplement at the beginning of this natural progression phase, where arrow 1 turns into arrow 2 and projects upward, the athlete will obviously associate it with the supplement and not natural progression. This natural progression phase is well understood by the supplement industry, which recognizes that all it must do is get the athlete to try its product just as this normal growth and development spurt occurs, and the athlete will associate the development to the product rather than just normal progression.

Variable 3—The Placebo Effect

When real expectations accompany the use of a product that promises certain mental or physical enhancements, there can be significant improvement in performance. These changes, however, are unrelated to any actual biological effects of the product. It's simply the placebo effect, the mental transformation of expectations into real gains, which will occur in a minimum of 30% of consumers exposed to affective marketing. In athletics, this placebo effect will be much higher.

As an example, let me use the product super-oxygenated water (SOW) as an example. The product(s) in this category claim SOW will enhance the performance of athletes by delivering more oxygen to the muscle tissue. Now if you are a goldfish, this product might be effective; if not, then any enhancement in performance is simply due to the placebo effect. Let me explain using a 2006 study on SOW by The American Council on Exercise (ACE), a nonprofit organization.[14] The study was conducted at the University of Wisconsin, La Crosse, using 32 health volunteers performing separate 5-km time trials. Here is the report from ACE:

> The ACE-commissioned study was led by Jennifer Otto, John P. Porcari, Ph.D., and Carl Foster, Ph.D., at the University of Wisconsin, La Crosse. The research team tested thirty-two healthy volunteers that represented both competitive and recreational runners who ran a minimum of 7.3 miles per week.
>
> Study participants were told that they were involved in a study to measure the effects of super-oxygenated water (SOW) on exercise performance. Each volunteer watched a short video detailing the purported beneficial effects of SOW and how their performance might be enhanced by drinking SOW before a race.

After preliminary tests to determine their fitness levels, each subject ran three separate non-paced 5-km time trials. The runs involved half the subjects drinking 16 ounces of bottled water or 16 ounces of what they thought was SOW (but was, in fact, tap water). During each trial, heart rate and rating of perceived exertion (RPE) were measured, while blood lactate concentration and running performance time were measured.

The results show that during the placebo trial, on average subjects ran 83 seconds faster, when they thought they were drinking SOW with 84 percent or 27 of 32 subjects running faster during the placebo trial. Heart rate, RPE and blood lactate levels were virtually the same between the two conditions.

"Over the years, placebo studies have shown that subjects who believe that they are receiving beneficial treatment often experience a variety positive outcome," said Dr. Cedric Bryant, ACE Chief Science Officer. "There clearly is a strong connection between the mind and body as it relates to physical performance."

Variable 4—Spiked Products With Stimulants and Steroids

Many athletes have a very naïve impression that the products they consume are inherently safe, which is hardly the case, as I point out in the liability of supplements, Chapter 9. The science literature, as well as the FDA's annual Adverse Events Report, clearly illustrate that many of these products can adversely affect the liver or kidney, and cause shortness of breath, tachycardia, chest pain, seizures, nausea, etc., to just name a few of the symptoms, as well as death. It is true that stimulants, as well as steroids, are going to enhance your workout and physical development, but at what price to your long-term or even your short-term health? The potential for a cardiovascular disaster or other health problem is very real, and to believe that the short-lived benefits to many of these products is related to the protein content is very naïve at best.

Here are a few examples from my own personal experience.

* A 22-year-old bodybuilder who ingested his pre-workout product at 7 a.m. At 9 a.m. he entered our clinic for a pre-employment screen. Resting blood pressure was 280/120 (he was told to immediately to go to the E.R. or see his M.D.). Twenty-four hours later, 150/80.
* 29-year-old male athlete who ingested two energy drinks, one at 4 a.m. and the other at 9 a.m. His resting blood pressure was 190/110 at 9:30 a.m. Twenty-four hours later, 130/85.
* 20-year-old college student taking my evening nutrition course. His senior year during track season, he suffered a stroke and a heart attack during a weight training session after consuming several over-the-counter pre-workout products.

- 28-year-old male drank his pre-workout drink at 6 a.m. Blood pressure at 10 a.m. was 170/100. Twenty-four hours later, 130/80.
- 30-year-old bodybuilder returning to work after being originally suspected as developing multiple sclerosis two years earlier, later recognized as B6 toxicity from the ingestion of a wide variety of sport supplements. He dropped the supplements, regained most of his lost motor skills, and was returning to work.
- A 39-year-old male bodybuilder who came in for a pre-employment screen for a local oil company at 4:20 p.m. He stated he had taken his pre-workout drink that morning at 10 a.m. His resting blood pressure was 219/135. Twenty-four hours later, 150/90.

The take-home point: time, persistence, good diet, and good coaching should be the focus of the athlete's training program, not potions with unknown ingredients with unknown effects on their health. The use of many supplement industry products is comparable to playing Russian roulette with your health, as stated in Chapter 9.

Misconception 6: How Much Muscle Development Can the Athlete Expect in One Week?

Supplement manufacturers exaggerate the *amount* of muscle mass that can be developed with the aid of protein supplements. The primary factors determining muscle development is genetics and time spent training. Obviously, diet, rest, well-designed training protocols, etc., play significant roles, but ultimately, any individual physical development will be determined by his or her genetics, unless steroids are utilized. Even with continual strenuous physical exercise, significant muscle gains after you reach your mid-twenties to early thirties are not likely for most people. Each year thereafter, the developmental rate slows to half—or less—of what it was to the point that development progresses at a snail's pace or stops completely, when age and genetic limitations are achieved, and muscle maintenance begins.

Steven J. Fleck, Ph.D., and William J. Kraemer, Ph.D., stated the following in their book *Designing Resistance Training Programs*:[15]

> The largest increases in lean body mass are a little greater than 3 kilograms (6.6 lb) in 10 weeks of training. This translates into a lean body mass of 0.66 pounds per week. Though some coaches desire huge gains in body weight for their athletes during the off-season, this is impossible if that added body weight is going to be muscles mass.[16]

Once you have a basic understanding of both the possibilities and limitations of muscle development, a mental "quack alarm" should sound as soon as you hear about a way to gain 2–3 pounds per week of solid muscle. You're obviously dealing with someone uneducated in muscle physiology and nutrition, someone who will stretch the truth to make money, or, more charitably, someone who has misinterpreted gains in fat mass, bone mass, increased water retention, or increased stored sugar as muscle mass development.

Muscle physiology is very efficient, enabling us to walk great distances and work long, physically challenging hours while still having the strength to perform other necessary tasks. If our bodies needed the enormous protein intake promoted by supplement manufacturers, there would be significant muscle wasting in athletes and blue-collar workers who do not consume high-protein diets or take protein supplements. Keep in mind that amino acids are readily recycled to assist in rebuilding the damaged tissue. This is an issue most athletes and active consumers are unaware of.

Misconception 7: Can a Normal Diet Provide Enough Protein?

Yes, and it is quite easy to do so. Most consumers already consume more protein than they need, and athletes, for the most part, are no exception. Consider the following options:

1. 3 cups nonfat milk (27g).
2. 3 cups rice and beans (27g).
3. 4 ounces chicken (32g).
4. 3 oz canned tuna in water (20g).
5. 16 oz low-fat plain yogurt (20g).
6. 3 oz lean beef or pork (20g).
7. 3 large eggs (20g).
8. 3 oz cheddar cheese (20g).
9. 1.5 oz cottage cheese (20g).

In 2000, *Nutrition in Sport Volume VII* of the *Encyclopedia of Sports Medicine*, an International Olympic Committee Medical Commission publication in collaboration with the International Federation of Sports Medicine, stated the following on page 163 under the heading "Protein Supplementation: Is It Necessary?": "Protein supplementation is probably not necessary for the vast majority of physically active individuals because the amounts of protein found to be necessary (1.2–1.8 g/kg/day) can be obtained in one's diet assuming total energy intake is adequate."[17]

An additional key factor here is to consume quality protein sources more often than the standard three meals per day for some athletes.

Let me provide a reasonable dietary assessment which anyone can who is attempting to build more muscle mass can do. For most, protein intake is normally not the issue, but if it is, it can be easily corrected with wiser protein sources throughout the day. The problem for many is going to be total calories if your training protocols are rigorous. Simply keep track of your total calorie intake throughout the week. Maintain this calorie intake for several weeks and monitor visible body fat levels in a mirror in your birthday suit. Starting on a Monday, add 250–500 additional calories to your daily diet and be consistent with the training load. Do this for one week. On Monday the following week, assess body fat levels visibly again. If there is no apparent increase in body fat, then increase your total caloric intake again another 250–500 calories per day. Continue with this process until you finally notice some accumulation of body fat. At this point, your body is illustrating to you that you are now consuming more calories than you need to support your workload and has begun to store the excess as fat mass. At this point, then simply reduce your daily energy intake back to the previous week, which should now represent the optimal energy intake you need to meet energy demand, maintain maximal natural testosterone levels, as well as maximal muscle development.

Notes

1. USDA Retail Milk Prices Report, *Agriculture Marketing Services* (December 24, 2018), p.1. https://www.ams.usda.gov/sites/default/files/media/RetailMilkPrices2018.pdf
2. P. Lemon, Effects of exercise on dietary protein requirements. *International Journal of Sport Nutrition* (1998), Vol. 8, pp. 426–47.
3. Janice Palmer, Too much protein can lead to dehydration, researchers find. *University of Connecticut Advance* (April 29, 2002). http://advance.uconn.edu/2002/020429/02042904.htm
4. T. Stokes, A.J. Hector, R.W. Morton, C. McGlory, S.M. Phillips. Recent perspectives regarding the role of dietary protein for the promotion of muscle hypertrophy with resistance exercise training. *Nutrients* (February 7, 2018), Vol.10, No. 2, p. 180.
5. R. Wolfe, Branched-chain amino acids and muscle protein synthesis in humans: myth or reality? *Journal of the International Society of Sports Nutrition* (2017) Vol. 14, No. 30, p. 5.
6. Jant Cholsoon, et al., A branched-chain amino acid metabolite drives vascular fatty acid transport and causes insulin resistance. *Nature Medicine* (2016) Vol. 22, pp. 421–6.
7. R. Wolfe, Branched-chain amino acids and muscle protein synthesis in humans: myth or reality? *Journal of the International Society of Sports Nutrition* (2017) Vol. 14, No. 30. p. 7.
8. Jant Cholsoon, et al., A branched-chain amino acid metabolite drives vascular fatty acid transport and causes insulin resistance. *Nature Medicine* (2016) Vol. 22, p. 1.
9. Michelle E. Baker, Kristen N. DeCesare, Abby Johnson, Kathleen S. Kress, Cynthia L. Inman & Edward P. Weiss, Short-term mediterranean diet improves endurance exercise performance: A randomized-sequence crossover trial, *Journal of the American College of Nutrition* (February 13, 2019),
10. Michael Gleeson, David C Nieman, and Bente K Pedersen, Exercise, nutrition and immune function. *Journal of Sports Sciences* (2004), Vol. 22, No. 1, pp. 115–25.
11. Ibid, p. 1 (abstract).
12. B. Pedersen, J. Helge, E. Richter et al., Training and natural immunity: Effects of diets rich in fat or carbohydrate. *European Journal of Applied Physiology* (2000), Vol. 82, No. 98. https://doi.org/10.1007/s004210050657
13. Michael Gleeson, David C Nieman, and Bente K Pedersen, Exercise, nutrition and immune function. *Journal of Sports Sciences* (2004), Vol. 22, No. 1, (abstract).

14. acefitness.org/about-ace/press-room/press-releases/415/ace-tests-mind-over-bodyexclusive-ace-research-tests-placebo-effect
15. William J. Kraemer and Steven J. Fleck, *Designing Resistance Training Programs*, Human Kinetics, 1987.
16. Ibid, p. 154.
17. https://epdf.pub/nutrition-in-sport.html

7

THREE REASONS SUPPLEMENTS WILL NOT BENEFIT MOST PEOPLE

According to a May 2019 report from Grand View Research (GVR), the global dietary supplement market will be "worth $194.63 billion by 2025."[1] GVR also stated:

- The market is backed by rising health awareness among consumers of all age groups across the world coupled with a considerable increase in the number of fitness centers and gymnasiums.
- North America emerged as the largest market for dietary supplements in 2018. On a macro level, rising awareness pertaining to nutritional enrichment expected to propel the regional demand in the forthcoming years. Moreover, rising demand of sports as an academic curriculum activity in education systems in major markets including the U.S., Russia, China, and Japan is expected to promote the application of dietary supplements among children.
- Energy and weight management is expected to remain dominant application segment throughout the forecast period owing to the rising awareness regarding fat reduction and focus on enhancing nutrition intake among adults.

This misguided spending illustrates that most consumers embrace a wide variety of myths and misconceptions regarding the purported benefits of most supplements, as well as a lack of understanding of three distinct biological mechanisms which help to maintain nutrient homeostasis (balance) over a broad range of intake. These misunderstandings make consumers easy to exploit and fleece by the supplement industry. These three homeostatic mechanisms are:

1. Reserve capacity of all nutrients.
2. Changes in absorption rate of all nutrients when needs increase.
3. Increased recycling rates when needs increase.

Chapter 9 illustrates the manufacturing problems, as well as the health and safety concerns associated with the supplement industry. Chapter 8 illustrates the antiquated reasoning for the use of antioxidant supplements. This chapter will illustrate why supplements cannot work in most cases, a fact that most consumers—as well as many professionals—are unfamiliar with.

This is not to say that some supplements are not appropriate for some users under various circumstances, as discussed at the end of this chapter, but most supplement users purchase them because they embrace some misunderstanding about the purported need or benefit for them.

The supplement industry and authors of many popular trade books on health and nutrition, as well as many misinformed professionals in the field, would have you believe that obtaining the appropriate quantities of vitamins and minerals is not possible without supplementation, often stating even if you adhere to good dietary habits. Now, is this true? Do the biological roles each nutrient is involved in become negatively affected if you fall short of the daily recommendation for the nutrient? Does a higher level of physical training elicit a greater need which cannot be obtained from food alone? Let's look at the evidence.

First, as I have stated elsewhere in this book, most consumers have a complete misunderstanding of the recommended intakes most developed countries have developed. In the United States, they are referred to as the Recommended Dietary Allowances (RDAs). Many believe that the RDAs will only meet their minimum requirements, and likely less if they are athletic or very active. I provided an illustration of this in Chapter 1 using the comments of a Major League Baseball strength coach. However, this is a false assumption. Alfred Harper, Ph.D., past professor Emeritus at the University of Wisconsin Department of Biochemistry and Nutritional Sciences who served on and chaired the Food and Nutrition Board (FNB) of the National Academy of Sciences for many years, as well serving as past president of the Federation of American Societies for Experimental Biology (FASEB), clarifies the purpose of the RDAs. "RDA for essential nutrients are not average requirements. They *exceed* [my emphasis] the needs of most, if not all, individuals in the specified groups, specified by age and sex."[2]

What does this mean in a practical sense? Let me illustrate using a national television advertisement for one of the largest over-the-counter multivitamin products in the United States. The product: One-A-Day Women, as well as the men's formula, "for nutritional support." The TV commercial called "Healthy Americans" for 2015 (Ad ID: 1176769) states the following: "Ninety-percent fall short in getting key nutrients from food alone." The advertisement states at the bottom of the screen that apparently, the data shows that most Americans only ingest 50% of the recommended amounts for vitamin A per day, 40% for vitamin C, 90% of vitamin D, 90% of vitamin E, and 50% of calcium. Therefore, the obvious takeaway from this advertisement is that the company wants consumers to believe that they have a significant need for a daily supplement because they are not meeting their purported daily requirements. Now let me illustrate using the three biological mechanisms mentioned previously that this advertisement is very deceptive and is simply exploiting typical misunderstandings of most consumers to sell supplements.

Mechanism 1: The Reserve Capacity of Nutrients

Consumers are often told that they need to consume supplements daily to sustain the maximum level of concentration in the tissues (saturation) and maintain optimal health. The problem is that there is no evidence that this concept is true. The saturation point, or maximum reserve capacity, is simply the upper end of retention for a given nutrient before it is excreted in the urine, catabolized to other products, or, possibly, turns toxic—*the Principle of Toxicology*.

The late Victor Herbert, M.D., J.D., was past professor of medicine at Mount Sinai School of Medicine and a FASEB recognized expert in nutrition. He was also Chief of Mount Sinai's Hematology and Nutrition Research Laboratory and former president of the American Society for Clinical Nutrition. He received the FDA Commissioner Special Citation in 1984 for "outstanding and consistent contributions against proliferation of nutrition quackery to the American consumer." He was also past Fellow of The American Society for Nutritional Sciences. Dr. Herbert states the following:

> Contrary to popular belief, it is not necessary to consume water-soluble vitamins every single day: the body stores enough to provide reserves that last for a few weeks to months or even longer in some cases. It is important, however, that the average intake over a week or two provides the variety of foods that supply all the essential vitamins. The liver is the body's main nutrient storehouse. It absorbs and stores excess nutrients from the blood, and releases into the blood those nutrients that are not coming into the blood from the diet.[3]

The Colorado State University Extension "Fact Sheet" on fat-soluble vitamins A, D, E, and K—9.315 states essentially the same thing. Under the "Quick Facts" section, it states, "the body does not need these vitamins every day and stores them in the liver and adipose (fat) tissue when not used."[4]

Do not misunderstand. Biochemical abnormalities can certainly be displayed before the reserve of a given nutrient is completely exhausted. But the point is that there is a large buffer zone between maximum reserve and what must be present for proper cellular function and development. Let me provide a few examples.

Copper

Some nutrients, such as copper, have relatively little reserve compared to other trace elements. Still, it is significant when compared to our daily needs. An adult human contains about 50–120mg, which is bound to an enzyme or protein molecule in the liver[5] but the daily need in an adult is only 1–2mg. Thus, even though copper reserve does not appear to be significant, it is more than adequate for times of reduced availability.

Vitamin C

An average-size man can have a total body pool of vitamin C as high as 1,500–2,000mg. Consuming 60mg/day of vitamin C can attain this level, and it would take approximately 60 days to decline to 300mg with negligible intake. Current daily recommended levels are 90mg for men and 75mg for women in the United States. According to Alfred Harper, Ph.D., who I introduced previously, an intake of "30mg per day, an amount that would maintain a body pool of about 1000mg and provide a reserve lasting 20 to 30 days, should be ample."[6] Thirty milligrams is available in one-half of an orange. Dr. Harper also stated:

> Differences in judgment on the size of the desirable reserve resulted in the FAO/WHO recommending 30mg per day of ascorbic acid as a safe intake for men, whereas the U.S. NRS/NAS recommends 60mg per day or more, implying that saturation of tissues should be the criterion of adequacy and that individuals consuming between 30 and 60mg daily are at risk for inadequate intakes, even though no evidence of the health benefit from the higher intake has been demonstrated.[7]

Despite these recommendations, the most recent RDAs recommend 90mg per day, even though the benefits are unproven. The Australian National Health and Medical Research Council reviewed the same data to arrive at their Nutrient Reference Values for Australia and New Zealand men and women, and set its recommendations at 45mg/day for both groups. The British Nutrition Foundation established 40mg/day for both men and women, as did Public Health England, all considerably lower than the US values.

However, the point is not which country's recommendations is correct, but that either 40mg or 90mg per day will maintain a substantial reserve of vitamin C, and how easy it is to obtain through food. How hard is this to maintain without vitamin supplements? Easy—very easy. Following is a list of common foods and their vitamin C content as taken from *Bowes and Church's Food Values of Portions Commonly Used*, 14th edition:[8]

1.   Cantaloupe, 1 cup = 68 mg.
2.   One medium guava = 165 mg.
3.   Kiwi fruit = 75mg.
4.   One navel orange = 80 mg.

5. Strawberries, 1 cup = 85 mg.
6. Frozen peaches, 1 cup = 235mg.
7. Red raw chilies (3½ oz) = 369 mg (if you can handle the heat).

The important point to remember here is that maintaining good vitamin C levels is not difficult.

Robert Jacobs, Ph.D., F.A.C.N., past research chemist with the USDA Western Human Nutrition Research Center at UC Davis, author of the vitamin C chapter in the college text *Modern Nutrition in Health and Disease 8th edition*, and a past panel member for the development of the RDA's, noted that "no signs of ascorbic acid deficiency have been shown at this pool size (900mg), and that higher body reserves have not been shown to provide increased health benefit."[9] Dr. Jacobs' comments are consistent with those of Dr. Harper's mentioned previously. Thus, despite widespread declarations that supplements are necessary to maintain maximum nutrient reserve, evidence clearly shows that optimal health can be maintained without saturating the nutrient reserve capacity. Even when intake is minimal the overall body pool turnover is roughly 3% a day, a fraction of what is in reserve.

Now let's compare this information against what the national advertising campaign for One-A-Day's multivitamin stated previously as their reasoning to take their product. For vitamin C, they support their claim because consumers purportedly only get 40% of the RDA. Is this a real concern, or a fabricated fear? Should it warrant a supplement?

The RDA in the United States is 90mg/day, which will maintain total saturation of reserve capacity, and then some. The 40% figure used in the advertisement to entice consumers is 36mg (40% of 90). This **would maintain a body pool of about 1,000mg** [my emphasis] and provide a reserve lasting 20–30 days, as stated previously by Dr. Harper. The 36mg, or 40% used in the One-A-Day supplement advertisement, would meet the daily requirements of most every other industrialized country daily recommendation. So, is the One-A-Day advertisement providing consumers with enough information to make an educated choice regarding their actual need for the product? Here is the analogy I have already used, but it continues to help clarify this point. You are driving down the road and an advertisement comes on the radio from a major fuel company, which states "the majority of drivers make their daily commutes with only 40% of their maximum fuel tank capacity, so stop at the next service station and fuel up to make sure your vehicle runs properly." How would you respond?

Vitamin E

Ronald Sokol, M.D., Section Head, Gastroenterology, Hepatology and Nutrition, University of Colorado School of Medicine, authored the chapter on vitamin E in the college postgraduate-level nutrition text *Present Knowledge in Nutrition 7th edition*. In this chapter, he points out that sufficient levels of vitamin E are easily maintained by eating a variety of foods, and a deficiency is rarely found:

- Symptomatic deficiency of vitamin E rarely, if ever, occurs in humans because of inadequate oral intake of the vitamin, probably due to its wide distribution in foods.[10]
- Even when malabsorption of vitamin E does occur in adults, it takes several years before plasma vitamin E levels decreased to a deficient range because of the presence of body stores.[11]

Phillip Furrell, M.D., Ph.D., from the University of Wisconsin, and Robert Roberts, M.D., of the University of Virginia, also comment upon the rarity of vitamin E deficiency in *Modern Nutrition in Health and Nutrition, 8th edition*:

- Rapid development of vitamin E deficiency in humans apparently does not occur except in unusual clinical circumstances.[12] The occurrence of vitamin E deficiency of pure dietary origin is rare in developed countries.
- Despite the evidence that healthy levels of vitamin E can easily be maintained, recently, the Food and Nutrition Board recommended that the RDA for vitamin E be increased to 15mg per day versus the last RDA of 10mg. They seem to ignore the fact that millions of Americans maintain excellent health on one-third of this amount.

In the 11th edition of *Modern Nutrition in Health and Disease*, Maret Traber, Ph.D., of Oregon State University states that vitamin E deficiency "occurs only rarely in humans and virtually never as a result of dietary deficiency."[13]

The British Nutrition Foundation states, "Existence of dietary vitamin E deficiency is not considered to be a problem even in people consuming a relatively poor diet. Deficiency only occurs in people with severe fat malabsorption and rare genetic disorders."[14]

To support these points, consider the comments of the late Max Horwitt, M.D. of the Saint Louis University School of Medicine, in his published critique of the requirement for vitamin E in the *American Journal of Clinical Nutrition*.[15] Dr. Horwitt was a member of three RDA committees and was involved in the field of vitamin E research for over 65 years.

1. The Food and Nutrition Board (FNB) recommendation "was largely based on the studies in the Elgin Project reported in 1960." He points out that he was involved in those studies, and states "that no specific requirement for vitamin E be adopted on the basis of these studies, despite my involvement with them." This is due to the long-term studies used by the FNB to justify the increase in the RDA for vitamin E utilized experimental diets, which "contained large amounts of oxidized unsaturated fats not found in habitual

diets." Additionally, Dr. Horwitt points out that in those individuals who received the same clinical supervision as the experimental group, but consumed a controlled hospital diet, demonstrated no pathological changes even though they consumed a diet low in fruits and vegetables that "provided less than 8mg vitamin E/day." This, he states, is why the 1989 RDA Committee "concluded that the requirement for men should be 10mg vitamin E per day although it was known that millions of persons have lived long lives while consuming much less. In addition, increasing this requirement to 15mg/day benefits only the commercial interests involved in the sale of vitamin E."

2. "Having been a member of 3 other RDA committees, I am aware of the extensive debate that occurs at each meeting. It is my judgment that more attention should have been paid to the amount of vitamin E consumed by millions of healthy individuals," which, as he points out, is about 1/2 of the 1989 RDA of 10mg for adult males, and 8mg for adult females. These reduced levels were found to be "without any known apparent harm to the subjects evaluated."

3. Supplements that provide more than "the amount greater than the RDA is a pharmacologic dose, not a nutritional requirement."

The points Dr. Horwitt makes all conclude that RDA recommendations for vitamin E intake are excessive. He shows that most individuals consume much less, while still maintaining good health. While exaggerating the needed level of daily intake of vitamin E may boost supplement sales, it will not help the supposed true focus of those supplements: your health.

Now recall what the national advertising campaign for One-A-Day's multivitamin stated as the reasoning to take their product. For vitamin E, they support their claim because consumers purportedly only get 90% of the RDA. The RDA in the United States is 15mg/day. So, 90% of 15mg/day is 13.5mg/day, which, as explained earlier, is far more than adequate. Other industrialized countries, such as Australia and New Zealand, recommended daily intake is just 10mg/day for men, the older US standard, and 7mg/day for women. So, is the One-A-Day commercial pointing out a real concern or a fabricated concern? Do the facts warrant a supplement or clarify the nonsense? If you already consume at least 13.5mg/day, 90% of the US inflated need for the nutrient, with no known health benefit even at that level of intake, should you really consider a supplement as suggested by the One-A-Day advertisement?

Vitamin D

Vitamin D is readily stored in the liver and fat cells, and when intake is low, both sites of storage can slowly release vitamin D to maintain normal blood levels of it for several months. Additionally, as pointed out by the Colorado State University "Fact Sheet" on fat-soluble vitamins, just "ten to fifteen minutes of sunlight without sunscreen on the hands, arms and face, twice a week is sufficient to receive enough vitamin D."[16]

Now let's consider again the One-A-Day advertisement which stated that 90% "fall short in getting key nutrients from food alone," and implied that obtaining 90% of the recommended daily intake of vitamin D from food is problematic. First, 90% of the recommended levels is more than enough to maintain normal physiological functions. Second, even if this were not true, any minor amount of sun exposure throughout the week would be far more than enough to fill the implied unhealthy gap of 90% and 100%, which would of course negate the need for any supplement. This last point illustrates the major irony of the One-A-Day advertisement. If you watch the segment, it will show literally hundreds of individuals outside enjoying various outdoor activities with ample skin exposure, which is allowing their skin to do what? Initiate the biochemical process which begins with the skin's exposure to sunlight and makes more than enough vitamin D. The advertisement should have simply stated, "if your diet only provides 90% of the recommended vitamin D, then simply go outside, run around, or sit in the sun."

Mechanism 2: Changes in Absorption Rates

Now that you recognize the role nutrient reserve capacity has in maintaining nutrient balance, you also need to understand the role of how increased absorption rates of each nutrient can be triggered during times of increased need or reduced reserve.

Most of the digestion and absorption of food and its nutrients into the bloodstream occurs in the small intestine. According to research results reported in 2014 in the *Scandinavian Journal of Gastroenterology* titled "Surface Area of the Digestive Tract—Revisited,"[17] the small intestines diameter is 1 inch or 2.5cm, its length is 10 feet or 3.05 meters. Researchers determined the unique projections and folding's of the villi and microvilli, which line the small intestines absorptive surface area, "amplify the small intestinal surface area by 60–120 times," illustrating the highly complex and efficient capacity of the small intestines to regulate the absorption of nutrients by two-fold to four-fold. This can vary dramatically, depending upon need, availability, and negative interactions with other nutrients taken in excess, which may impede a nutrients absorption. As an example, large quantities of zinc, such as from a supplement, can interfere with copper bioavailability.[18]

The authors of one research study showed that just "an intake of zinc only 3.5 mg/day above the RDA for men reduced apparent retention of copper."[19] This means that if you ingest a large supplemental dose of zinc with the belief that it may have some special health attributes, it may end up precipitating lowered copper availability.

However, this section is not about what restricts nutrient absorption, but what enhances absorption. Specifically, when nutrient needs increase, absorption rates increase to sustain normal physiological activity. Here are a few examples.

Copper

According to James Collins, Ph.D., Food Science and Human Nutrition Department, University of Florida, "absorption of dietary copper is regulated, with the percentage of absorption increasing when intakes are low."[20] Under normal conditions copper absorption is roughly 10%. This leaves plenty of latitude for absorption rate increases when necessary. As an example, in July 1998, in the *American Journal of Clinical Nutrition*, the study "Copper Absorption, Excretion, and Retention by Young Men Consuming Low Dietary Copper Determined by Using the Stable Isotope 65Cu," was published. It was conducted by Judith Turnlund, Ph.D., Research Nutrition Scientist with the USDA Western Human Nutrition Research Center in San Francisco at the time.

Eleven young men were confined to a metabolic research unit for 90 days. The study found that young men could maintain their copper status for 42 days with a dietary copper intake of slightly less than 0.8mg/day. This intake is below the current Estimated Safe and Adequate Dietary Intake (ESADDI) of 1.5–3mg/day. It seems that humans adapt to different amounts of dietary copper by varying the efficiency of copper absorption as well as excretion. When dietary copper was 0.8mg/day, absorption was 56%, but when the intake was 1.7mg/day, absorption declined to 36%. When intake was 7.5mg/day, copper absorption was 12%.[21]

In her chapter on copper metabolism in the text *Present Knowledge in Nutrition*, Maria Linder, Ph.D., Department of Chemistry and Biochemistry, California State University at Fullerton, says that "there appears to be an adaptation to levels of intake, so that there was a greater efficiency of absorption at lower intakes, and vice versa."[22] Noting that the actual need for copper may be less than 1mg/day in adults, she said there is little evidence of copper deficiency within the populations of industrialized nations.

Calcium

When individuals went from 2,000mg/day of calcium to 300mg/day, absorption rates increased by 66% for women, age 22–31 years and 50% for women, age 61–75 years.[23]

Mechanism 3: Retention or Changes in Excretion Rates

As noted earlier, many people believe that nutrients are used up and then what is left is quickly excreted or catabolized. This is a myth. Specifically, nutrient excretion rates through the kidneys readily change from rapid (when intakes are excessive) to very low (when intake is low or needs increase). The following examples demonstrate how dramatic changes in excretion, and the consequential reabsorption rates, can play a significant role in maintaining nutrient balance over a wide range of dietary intakes.

Calcium

Lindsay Allen, Ph.D., R.D., professor of nutritional sciences at the University of Connecticut, and Richard Wood, Ph.D., from the USDA Human Nutrition Research Center on Aging at Tufts University, state that the body quickly adapts itself to varying intake levels, and that sometimes, less is more. "When calcium intake in human subjects is reduced abruptly from a high or adequate to a low level," they write, "within 1 week there is an increase in serum parathyroid hormone and active vitamin D (to enhance absorption), an increase in the fractional retention of calcium of about 50%, and a reduction in urinary calcium."[24] This mechanism is effective to not only compensate during times of low intake, but also when there is an increase in need, as in pregnant and lactating women, and growing children.

Vitamin C

When intake of vitamin C is less than 100mg, no ascorbate is excreted in the urine. At 100mg, about a quarter of the dose is excreted, and at 200mg, the figure rises to half. By contrast, for doses greater than or equal to 500mg, all the absorbed dose is excreted in urine. At higher doses, virtually all the absorbed dose was excreted.[25] Thus, despite an individual's best intentions to supplement vitamin intake, the body naturally excretes what it does not need, resulting in a waste of money.

Manganese

John Finley, Ph.D., with the USDA Agricultural Research Service, studied the absorption and retention rates of manganese among 26 healthy, nonpregnant women, ages 20–45.[26] The results again show the body's ability to adapt excretion levels to the body's current needs:

- Despite large differences in intake, the body effectively controlled ultimate retention of manganese.
- Calculated percentage retention of radioactive manganese after 60 days was five-fold to ten-fold greater in those eating the low manganese diet than those in the high diet.
- Excretion is an important means of controlling the balance of manganese, so excretion rates declined dramatically to compensate for reduced intake.

Copper

Judith Turnland, Ph.D., stated that humans excrete little during copper deficiency or when dietary copper is low. "The homeostatic regulation of copper absorption and excretion protects against copper deficiency and toxicity over a broad range of dietary intakes."[27]

Selenium

Homeostasis of selenium in the body is achieved through regulation of excretion. As dietary intake increases from the deficient range into the adequate range, urinary excretion of the element increases and accounts for maintenance of homeostasis . . . urinary excretion is the primary means whereby body selenium is regulated.[28]

Magnesium

When magnesium intake is suddenly and severely restricted in humans with normal kidney function, magnesium output (kidneys) becomes very small within 5–7 days. Supplementing usual dietary intake increases urinary excretion without significantly altering serum concentrations provided that renal function is normal, and the amount given does not exceed maximum filtration and excretion capacities. *The intestinal and renal absorptive and excretory mechanisms in normal individuals permit homeostasis to occur over a wide range of intake* [my emphasis].[29]

Final Comments

The most recently emphasized comment condenses the point of this chapter into one sentence. The literature could not be any clearer, apart from excessive iron losses in premenopausal women, which often requires the need for an iron supplement, the vast majority of nonpregnant individuals who consume supplements, believing they will benefit from them, have been duped. The body naturally adapts to low or high nutrient intake levels to a certain extent, and either by relying on nutrient reserve, or adjusting rates of absorption and excretion, it can self-regulate to achieve nutrient homeostasis and normal physiological processes within a broad range of intake. This information should counter the deceptive advertising used by the supplement industry, which would like you to believe that if you are not consuming 100% of daily recommended levels, you are at risk for metabolic abnormalities. This is demonstrably false. However, these three mechanisms are certainly not without their limitations. Millions of people around the world certainly struggle with malnutrition issues, as well as many in industrialized countries, because of poverty, neglect, poor lifestyle habits, ignorance, or by choice. However, for the average household in America, or any industrialized country, with its vast quantities of healthy food available, there is no reason not to be well nourished, vs. just overfed, with no need of a supplement—with some exceptions.

Moreover, bear in mind that poor food choices also eliminate the thousands of other plant chemicals (phytochemicals) which play just as important a role in long-term health as the "major nutrients" available in supplements. It is extremely unwise to assume that anyone has enough understanding of the complex effects on health of the various plant chemicals, to forgo personal responsibility to eat well, rather than rely on purported "health" concoctions. *You cannot reproduce what you do not have the blueprints for.*

Notes

1. www.grandviewresearch.com/press-release/global-dietary-supplements-market.
2. Maurice E. Shils, James A. Olson, and Moshe Shike, *Modern Nutrition in Health and Disease* (MNHD), 8th ed. (Philadelphia, PA: Lea & Febiger, 1994), p. 1475.
3. Victor Herbert, *Total Nutrition: The Only Guide You'll Ever Need* (New York, NY: St. Martin's Griffin, 1995), p. 99.
4. J. Clifford and A. Kozil, Fact Sheet on fat-soluble vitamins A, D, E, and K—9.315. *Colorado State University Extension* (September 2017), p. 1. https://extension.colostate.edu/topic-areas/nutrition-food-safety-health/fat-soluble-vitamins-a-d-e-and-k-9-315/
5. A. Catherine Ross, Benjamin Caballero, Robert Cousins, Katherine Tucker, and Thomas Ziegler, *Modern Nutrition in Health and Disease* (MNHD), 11th ed. (Wolters Kluwer Lippincott Williams & Wilkins), p. 211.
6. Maurice E. Shils, James A. Olson, and Moshe Shike, *Modern Nutrition in Health and Disease* (MNHD), 8th ed. (Philadelphia, PA: Lea & Febiger, 1994), p. 1480.
7. Ibid, p. 1480.
8. Jean Pennington and Helen Church, *Bowes and Church's "Food Values of Portions Commonly Used,"*(J.B. Lippincott Company), 14th ed., Philadelphia, p. 74–9.
9. Maurice E. Shils, James A. Olson, and Moshe Shike, *Modern Nutrition in Health and Disease* (MNHD), 8th ed. (Philadelphia, PA: Lea & Febiger, 1994), p. 473.
10. Ekhard E. Ziegler and L.J. Filer, Jr., eds., *Present Knowledge in Nutrition* (PKN), 7th ed. (Washington, DC: ILSI Press, 1996), p. 132.
11. Ibid, p. 133.
12. *MNHD*, 8th ed., p. 335.
13. *MNHD*, 11th ed., p. 300.
14. nutrition.org.uk/nutritionscience/nutrients-food-and-ingredients/vitamins.html?start=4, p. 5.
15. Max K Horwitt, Critique of the requirement for vitamin E, *The American Journal of Clinical Nutrition* (June 2001),Vol. 73, No. 6, pp. 1003–5.
16. J. Clifford and A. Kozil, Fact Sheet on fat-soluble vitamins A, D, E, and K—9.315, *Colorado State University Extension by* (September 2017), p. 1. https://extension.colostate.edu/topic-areas/nutrition-food-safety-health/fat-soluble-vitamins-a-d-e-and-k-9-315/
17. Herber Helander, and Lars Fändriks, Surface area of the digestive tract—Revisited. *Scandinavian journal of gastroenterology* (2014), p. 49.
18. *MNHD*, 8th ed., p. 216.
19. M.D. Festa, H.L. Anderson, R.P. Dowdy, M.R Ellersieck, Effect of zinc intake on copper excretion and retention in men, *The American Journal of Clinical Nutrition* (February 1985),Vol. 41, No. 2, pp. 285–92.
20. *MNHD*, 11th ed., p. 211.
21. J.R. Turnlund, W.R. Keyes, G.L Peiffer, and K.C. Scott, Copper absorption, excretion, and retention by young men consuming low dietary copper determined by using the stable isotope 65Cu. *American Journal of Clinical Nutrition* (June 1998),Vol. 67, No. 6. pp. 1219–25.
22. *PKN*, 7th ed., p. 315.
23. *MNHD*, 8th ed., p. 152.
24. Ibid, p. 152.
25. *PKN*, 7th ed., p. 152.

26. John Finley, Manganese absorption and retention by young women is associated with serum ferritin concentration, *American Journal of Clinical Nutrition* (1999), Vol. 70, pp. 37–43.

27. *MNHD*, 8th ed., p. 234.

28. *MNHD*, 11th ed., pp. 228–29.

29. *MNHD*, 8th ed., p. 171.

8

ANTIOXIDANT SUPPLEMENTS

Another Magic Bullet, or False Icon for Better Health and Performance?

The term antioxidant has become one of the holy grails for marketers of a wide variety of food and supplemental products. For many, as with the term "organic," the term "antioxidant" has become synonymous with better health and the prevention of a myriad of ailments. In essence, antioxidants have been misconceived to become the magic bullet for health and longevity. Antioxidants are chemical compounds which can be a vitamin, mineral, enzyme, or one of thousands of other naturally occurring plant chemicals, as well as many which are naturally produced by our bodies. One of their roles is to inhibit or stop potentially harmful oxidation reactions produced by free radicals.

As an example, Image 8.1 illustrates a free radical on the right with its missing electron at the top left of the molecule. This missing electron makes the free radical unstable and very reactive, both in a positive way as well as a possible negative way as explained ahead. However, if the molecule on the left is an antioxidant, the antioxidant will donate one of its electrons to the free radical, stabilizing the free radical, while continuing to remain stable itself. This prevents any further reactions of the free radical from continuing. This is a continuous biological process and a normal part of every cell's everyday existence, and there must be a balance of this reaction within the cell, which antioxidant supplements prevent.

For the general population, antioxidants are still promoted for antiaging. However, this is an old myth which persists today, even though most researchers in this area have long abandoned this misguided concept. As an example, the position statement on human aging published by *Scientific American* on May 13, 2002, entitled "The Truth About Human Aging," states that:[1]

> the purpose of this document is to warn the public against the use of ineffective and potentially harmful antiaging interventions and to provide a

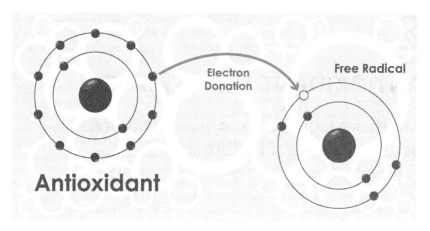

IMAGE 8.1 Chemical reaction illustrating an antioxidant donating an electron to stabilize a free radical.

brief but authoritative consensus statement from 51 internationally recognized scientists in the field about what we know and do not know about intervening in human aging.[2]

In the "Antioxidants" section, this group reported the following:

The scientifically respected free-radical theory of aging serves as a basis for the prominent role that antioxidants have in the antiaging movement. The claim that ingesting supplements containing antioxidants can influence aging is often used to sell antiaging formulations. The logic used by their proponents reflects a misunderstanding of how cells detect and repair the damage caused by free radicals and the important role that free radicals play in normal physiological processes (such as immune response and cell communication). Nevertheless, there is little doubt that ingesting fruits and vegetables (which contain antioxidants) can reduce the risk of having various age-associated diseases, such as cancer, heart disease, macular degeneration and cataracts. At present there is relatively little evidence from human studies that supplements containing antioxidants lead to a reduction in either the risk of these conditions or the rate of aging ... the possible adverse effects of single-dose supplements, such as beta-carotene, caution against their indiscriminate use..., but so far there is no scientific evidence to justify the claim that they have any effect on human aging.[3]

Within athletics or the physically active consumer population groups, antioxidants' main promotional hype is for the prevention of post-exercise muscle tissue damage, and the enhancement of muscle recovery through the reduction of free radical production. This is due to the alleged adverse effects free radicals are purported to have physiologically. For this theory to be true, the following four misconceptions must be true, which they are not.

1. "Free radicals are an abnormal part of cellular metabolism." False. Free radicals are a normal and necessary part of cell metabolism both at rest and during exercise.
2. "Supplemental antioxidants will enhance physical development and muscle recovery." False. Supplemental antioxidants may hinder physical development and muscle recovery.
3. "You are unable to obtain enough quantities of antioxidants from food." False. Food provides an overabundance of likely hundreds of antioxidants.
4. "Antioxidants are directly correlated to better health, so they must be beneficial as supplements." False. Antioxidant-rich foods are *associated* [my emphasis] with better health, but this benefit has nothing to do with the intake of these nutrients in an isolated form. The benefit comes from the synergistic effect of all the phytochemicals in those foods taken together.

In the 1990s, the free radical theory of aging and tissue damage was popularized based upon a limited understanding of the biochemistry of free radicals. At the time, knowledge of their positive roles in cellular metabolism and adaptation to stress was limited at best. This has dramatically changed the last 20 years, which clearly illustrates that free radicals are not only a necessary part of normal cellular metabolism, but the suppression of them can be harmful. The following review of these four misconceptions should clarify why consumers should refrain from ingesting supplemental antioxidants.

Misconception 1: Free Radicals Are an Abnormal Part of Cellular Metabolism

Advertisers of antioxidants do a wonderful job of instilling fear in consumers by directly associating free radicals with the development of chronic diseases, aging, or muscle tissue damage. This leads millions of consumers to rationalize, "why not load up on antioxidants and get rid of as many free radicals as possible?" However, this is a simple-minded approach to a very complex process. Free radicals are not an abnormal part of cellular metabolism. They are, in fact, a normal byproduct of many biological processes, and there must be a balance between normal and excessive production of them to maintain good health, which is referred to as cellular homeostasis. They are part of our cell's everyday life and have been unnecessarily demonized. Below are some examples of the positive biological roles of free radicals:

• Immune cells use free radicals to defend the body against viruses, bacteria and fungi, which is an essential immune defensive mechanism.
• Free radicals are essential signaling molecules between cells, regulating physiological processes. As an example, when muscle tissue has been significantly

stressed, as during strenuous exercise, free radicals act as a signaling mechanism to activate the physiological processes to enable the muscle tissue to adapt to the exercise. Without free radicals, muscle tissue adaptation may be inhibited.

- Free radicals help regulation of blood pressure.
- Free radicals cause cancer cells to "commit suicide," or what is biologically referred to as apoptosis, or programmed cell death. Suppression of this function may result in the unwanted proliferation of cancer cells.
- Oxidative stress (exercise) stimulates increased red blood cell production.
- Oxidative stress increases the production of the cell's energy producing organelle, the mitochondria, which are critical for all athletes, especially endurance athletes.
- Free radicals trigger the increased production of the cell's own protective antioxidants.

The Main Take-Home Message From Misconception 1

Free radicals are a normal byproduct of healthy cellular activity and should not be suppressed. Cells maintain a precise balance of oxidizing (free radicals) and antioxidizing reactions to stay healthy. Free radicals have been mistakenly demonized due to the lack of understanding of their important biological roles.

Misconception 2: Supplemental Antioxidants Will Enhance Physical Development and Muscle Recovery

Athletes and health-conscious consumers are continuously looking for the magic bullet to enhance their training and development. There are no magic bullets—just sweat and substantial effort. In fact, the misguided expectations of supplementing the diet with antioxidants may inhibit physical development and muscle recovery based upon studies conducted over the last ten years, as illustrated here.

Following are excerpts from various research studies conducted since 2006, which continue to support my position taken in *Muscles Speed and Lies*, published in 2006 and now out of print:

March 1, 2019—*International Journal of Exercise Science*, **"Antioxidant Supplementation Impairs Changes in Body Composition Induced by Strength training in Young Women."**[4]

The purpose of this study was to analyze chronic effects of strength training combined with vitamin C and E supplementation on body composition of college women. The main finding was that only placebo showed fat-free mass adaptations of greater magnitude ($P < .05$) when compared to control. Also, placebo was the only group that presented a significant decrease in fat mass after ten weeks of strength compared to baseline values.

In summary, the present results show that vitamin C and E supplementation attenuate (diminish) body composition adaptations related to strength training, as only participants that received placebo pills presented alterations of greater magnitude when compared to control. Effect sizes support this assumption by demonstrating higher values for placebo. *Considering these results, vitamin C and E supplementation should be avoided by healthy young women who want to increase FFM and reduce FM* [my emphasis].

March 9, 2018—*Cochrane Review*, "Antioxidants for Preventing and Reducing Muscle Soreness After Exercise."[5]

Our review included 50 randomized controlled trials with a total of 1,089 participants; 88% of the participants were men; 12% were women and the age ranged from 16 to 55 years. The antioxidants that were used varied from vitamin supplements such as vitamins C and E; whole foods such as blueberry smoothies and pomegranate juice; tea and extracts such as curcumin which is found in turmeric. The doses consumed were higher than the recommended daily amount, and the type of exercise undertaken also varied but was enough to induce muscle soreness.

On top of our findings, there has also been an emergence of studies showing that chronic antioxidant supplementation may actually be counterproductive. For instance, it has been shown that antioxidant supplements may delay healing and recovery from exercise and hinder adaptations to training, and may even increase mortality.

Taking all of this into consideration, antioxidants supplements are perhaps a waste of money. Instead, move more; exercise regularly; eat a balanced diet; one that includes at least five or more portions of rainbow-colored fruits and vegetables. Because for now at least, there appears to be no quick fix to ease muscle soreness after exercise. In fact, it seems muscle soreness is actually an important part of the recovery process and helps to make your muscles stronger and bigger over time. And it is this important recovery process that will ultimately help to make you fitter and stronger in the long run.

August 2018—*Journal of Applied Physiology, Nutrition, and Metabolism*, "Effect of Strength Training Combined With Antioxidant Supplementation on Muscular Performance."[6]

This was a placebo-controlled randomized study that aimed to investigate the effects of strength training combined with antioxidant supplementation on muscle performance and thickness. The study included 42 women ages 20–27 years old. The vitamin group was supplemented with vitamin C (1g/day) and E (400IU/day) during the strength training period. The results: "chronic antioxidant supplementation may attenuate peak torque and total work improvement in young women after 10 weeks of strength training."

July 19, 2016—*ScienceDaily* report, "Beware of Antioxidant Supplements, Warns Scientific Review."[7]

The lay press and thousands of nutritional products warn of oxygen radicals or oxidative stress and suggest taking so-called antioxidants to prevent or cure disease. Researchers have analyzed the evidence behind this. The result is a clear warning: do not take these supplements unless a clear deficiency is diagnosed by a healthcare professional.

July 28, 2016—*ScienceDaily* report, "Novel Drug Therapy Kills Pancreatic Cells by Reducing Levels of Antioxidants."[8]

While it's a matter of conventional wisdom in popular culture that raising antioxidant levels in the body tends to keep cancer at bay, a research team demonstrates that in pancreatic cells on the road to cancer or already malignant, the last thing one wants to do is to raise antioxidant levels. Lowering antioxidants in pancreatic cancer cells can help kill them, the study shows.

December 7, 2015—*The Journal of Physiology*, "Do Antioxidant Supplements Interfere With Skeletal Muscle Adaptation to Exercise Training?"[9]

Antioxidant supplementation has been more consistently reported to have deleterious effects on the response to overload stress and high-intensity training, suggesting that remodeling of skeletal muscle following resistance and high-intensity exercise is more dependent on ROS (free radicals) signaling. Importantly, there is no convincing evidence to suggest that antioxidant supplementation enhances exercise-training adaptations.

March 13, 2015—*The Physician and Sports Medicine Journal*, "Interplay of Oxidants and Antioxidants During Exercise: Implications for Muscle Health."[10]

In abstract: Free radicals trigger the "increase gene expression of antioxidant enzymes, cell protective proteins and other enzymes involved in muscle metabolic functions."

July 10, 2014—*ScienceDaily* report, "How Antioxidants Can Accelerate Cancers, and Why They Don't Protect Against Them."[11]

Two cancer researchers propose why antioxidant supplements might not be working to reduce cancer development, and why they may actually do more harm than good. Their insights are based on recent advances in the understanding of the *system in our cells that establishes a natural balance between oxidizing and anti-oxidizing compounds* [my emphasis]. These compounds are involved in so-called redox (reduction and oxidation) reactions essential to cellular chemistry.

December 2, 2013—*The Journal of Physiology*,**"Attenuated Effects of Exercise With Antioxidant Supplement: Too Much of a Good Thing?"**[12]

To summarize, eight weeks of exercise training enhanced several beneficial cardiovascular effects in aged men, but surprisingly, resveratrol (as an antioxidant supplement) attenuated many of these improvements (e.g. blood pressure and lipids, altered balance between prostanoid vasodilators and vasoconstrictors, and lower aerobic capacity). These effects contradict animal studies, suggesting the importance of rigorous controlled research in humans, *and, importantly, remind us of how vast differences between species can be expressed.* Therefore, ROS (free radicals) formation alone is not detrimental to cardiovascular health when appropriately regulated *endogenously* (e.g. following an appropriate exercise training program), but *exogenous* molecules (supplements) able to alter this balance could prove to be detrimental.

February 15, 2012—*American Journal of Physiology and Endocrinology and Metabolism*, **"Antioxidant Supplements in Exercise: Worse Than Useless?"**[13]

A significant number of both healthy and sick individuals are taking antioxidant supplements in the belief that these will improve their health and prevent or ameliorate diseases. Moreover, a large proportion of athletes, including elite athletes, take vitamin supplements, often large doses, seeking beneficial effects on performance. The complete lack of any positive effect of antioxidant supplementation on physiological and biochemical outcomes consistently found in human and animal studies raises questions about the validity of using oral antioxidant supplementation in both health and disease. Also, the vast majority of experimental evidence clearly advises against this supplementation. In our opinion, antioxidant supplements are, at the least, useless.

December 1, 2011—*Sports Medicine* **review, "Antioxidant Supplementation During Exercise Training: Beneficial or Detrimental?"**[14]

A growing body of evidence indicates detrimental effects of antioxidant supplementation on the health and performance benefits of exercise training. Indeed, although ROS are associated with harmful biological events, they are also essential to the development and optimal function of every cell.

The main findings of these studies are that, in certain situations, loading the cell with high doses of antioxidants leads to a blunting of the positive effects of exercise training and interferes with important ROS-mediated physiological processes, such as vasodilation and insulin signaling. We recommend that an adequate intake of vitamins and minerals through a varied and balanced diet remains the best approach to maintain the optimal antioxidant status in exercising individuals.

November 24, 2011—*Science-Based Medicine* review, "Antioxidants and Exercise: More Harm Than Good?"[15]

There is no strong evidence to suggest supplements have any meaningful effect on muscle damage. What's more concerning is that some studies have suggested that supplements may induce muscle injury and actually delay recovery. Also, supplements may reduce the ability to adapt to exercise-induced oxidative stress: cells will naturally adapt to increases in free radical production by upregulating (producing more) endogenous enzyme systems. Supplements may inhibit this endogenous adaptation. Finally, in light of what we know about antioxidants and exercise, the trend in the data is strongly suggestive of zero benefit, at best, with the real possibility that there may be negative consequences to supplementation. Overlay the epidemiology evidence that looks at mortality, cancer, and other outcomes, and the attractiveness of antioxidant supplements drops even further. The best advice for those who exercise seems to be to focus on consuming a diet rich in fruits and vegetables and leaving the antioxidant bottles on the shelf. There appears to be little that is complementary about them.

July 15, 2011—*Free Radical Biology and Medicine*, "Extending Life Span by Increasing Oxidative Stress."[16]

Abstract: We here summarize findings that antioxidant supplements that prevent these ROS (free radicals) signals from interfering with the health-promoting capabilities of physical exercise. Taken together and consistent with ample published evidence, the findings summarized here question Harman's Free Radical Theory of Aging and rather suggest that ROS act as essential signaling molecules to promote metabolic health and longevity.

July 2010—*Journal of Medicine and Science in Sports and Exercise*, "Antioxidant Supplementation Does Not Alter Endurance Training Adaptation."[17]

Our results suggest that administration of vitamins C and E (antioxidants) to individuals with no previous vitamin deficiencies has no effect on physical adaptations to strenuous endurance training.

July–August 2010—*Journal of Oxidative Medicine and Cellular Longevity*, "Exogenous Antioxidants—Double-Edge Swords in Cellular Redox State—Health Beneficial Effects at Physiological Doses Versus Deleterious Effects at High Doses."[18]

Abstract: High doses of isolated compounds may be toxic, owing to pro-oxidative effects at high concentrations or their potential to react with beneficial concentrations of ROS (free radicals) normally present at physiological conditions that are required for optimal cellular functioning.

December 2009—*Journal of Sports Medicine*, "Does Antioxidant Vitamin Supplementation Protect Against Damage?"[19]

There is little evidence to support a role for vitamin C and/or vitamin E in protecting against muscle damage. Indeed, antioxidant supplementation may interfere with the cellular signaling functions of ROS (free radicals), thereby adversely affecting muscle performance. Furthermore, recent studies have cast doubt on the benign effects of long-term, high-dosage antioxidant supplementation. Since the potential for long-term harm does exist, the casual use of high doses of antioxidants by athletes and others should be curtailed.

January 2008—*American Journal of Clinical Nutrition*, "Oral Administration of Vitamin C Decreases Muscle Mitochondria Biogenesis and Hampers Training—Induced Adaptations in Endurance Performance."[20]

Supplemental antioxidants appear to inhibit cellular adaptations to exercise. The second major conclusion which can be drawn from our experiments is that supplementation with vitamin C lowers training efficiency. The common practice of taking vitamin C supplements during training (for both health-related and performance related physical fitness) should be seriously questioned.

February 2007—*Journal of American College of Sports Medicine*, "Oxidative Stress in Half and Full Ironman Triathletes."[21]

Users of antioxidant supplements in both the half and full Ironman races had significantly greater evidence of oxidative stress (damage) after races compared with nonsupplementers.

The Main Take-Home Messages From Misconception Two

Supplemental antioxidants, at best, will have a neutral effect on your health, but may hinder natural cellular anticarcinogenic activity by inhibiting the normal healthy functions of free radicals. Additionally, supplemental antioxidants are just as likely to hinder muscle tissue adaptations to training through the inhibition of normal cellular and muscular adaptations, which would have been triggered by the production of free radicals now suppressed by the supplemental antioxidants. This does not occur with antioxidant-rich foods.

These studies should clarify why the typically marketed antioxidants provide no additional biological benefits, and, in some cases, *the Principle of Toxicology—the dose makes the poison*, should be considered. Is it even logical to flood the cell with higher-than-normal levels of a chemical compound which it already has enough quantities of? All chemical compounds, even vitamins and minerals, have an upper limit of safety.

Misconception 3: You Are Unable to Obtain Enough Quantities of Antioxidants From Food

This is a blatant fallacy. It is actually very difficult to become antioxidant deficient due to their widespread availability in food, as well as extensive reserve in the body. Consider the following points from several graduate level nutrition texts, *Present Knowledge in Nutrition* (PKN), 7th edition, and *Modern Nutrition in Health and Disease* (MNHD), 8th or 11th editions.[22, 23, 24]

Vitamin E

- "Symptomatic deficiency of vitamin E rarely, if ever, occurs in humans because of inadequate oral intake of the vitamin" (PKN 7th edition, page 132).
- Vitamin E deficiency "occurs rarely in humans and **virtually never** [emphasis mine] as a result of dietary deficiencies" (MNHD 11th edition, page 300).
- "Most ingested vitamin E, because of its relatively low intestinal absorption, is excreted in the feces" (MNHD 11th edition, page 298).
- "When malabsorption of vitamin E occurs in adults, it takes several years before plasma vitamin E levels decrease to a deficient range, because of the vitamin's presence in body stores" (PKN 7th edition, page 133).

Vitamin C

- "An intake of 30mg/day, an amount that would maintain a body pool of about 1000 mg and provide a reserve lasting 20 to 30 days should be ample" (MNHD 8th edition, page 1480).
- "At 60 mg/day or more [one orange], the body pool will approach the maximum size (saturation) of about 1500mg and will not fall to the point at which signs of scurvy develop (300mg) for at least 60 days, even if intake is negligible" (MNHD 8th edition, page 1480).
- "Vitamin C content of neutrophils, monocytes, and lymphocytes [your white blood cells for fighting infections] become saturated at a 100mg dose" (PKN 7th edition, page 151).
- The kidneys also readily absorb vitamin C.

Vitamin A

- The current RDA for vitamin A allows for "the safe level of intake which ensures adequate body reserves to meet needs for approximately 4 months on a diet low in vitamin A and during periods of stress, such as fever or diarrhea" (MNHD 8th edition, page 301).
- "Vitamin A is extensively stored in the liver and depleted from its reserves at a relatively low net rate of 0.5%/day" (PKN 7th edition, page 111).
- The average total body content of vitamin A would fulfill all functions of the vitamin and provide a 3-month reserve on a low vitamin intake (PKN 7th edition, page 114).

Beta-carotene

- "Among the 600 or more carotenoids that exist in nature, 50 show provitamin A activity" (MNHD 8th edition, page 290).
- "Carotenoids that are not converted into vitamin A still show significant biological and pharmacological activity" (MNHD 8th edition, page 290).

The reader can refer to Chapter 7, regarding the three biological mechanisms which help maintain nutrient homeostasis (balance) over a broad range of intake, for the reasons why supplemental antioxidants are a waste of money and time. Overly saturating cells with compounds which they need in only small amounts is counterproductive to the cell's normal metabolism and homeostatic mechanisms.

Misconception 4: Antioxidants Are Directly Correlated to Better Health, So They Must Be Beneficial as Supplements

It is a common misunderstanding that just because research illustrates a direct correlation between antioxidant-rich foods and a myriad of health benefits, then it must be the antioxidants which are providing the benefit. This is false. Let me provide an illustration. You are standing outside of a locked room waiting to get inside. The door opens, and ten individuals walk out. You then enter the room and find someone dead on the floor who has clearly been murdered. Which one of the ten individuals (variables), or combination of them, was responsible? They are all correlated, but as you can see, it is virtually impossible to determine which one of the variables, or combination of them, is the actual cause. Correlations do not indicate causation, only a possible variable, in the mix of variables, which may be the contributing factor. Likewise, it is well established that those who suffer less from a myriad of lifestyle-related diseases, and have improved physical function, consume diets which are naturally high in the various popular supplemental antioxidants. This is true. However, the antioxidants are only reflective of good dietary habits in general. They are just one variable (nutrient) in a mix of literally thousands of variables (other plant chemicals or nutrients), which are contained in the food. So, cherry-picking the antioxidants (variable) from the literally thousands of other possible plant chemicals (variables) as the one creating the positive impact is irrational at best. But this is what the supplement industry, as well as most marketers of many packaged food items and drinks which contain antioxidants, would like you to believe. Instead of labeling antioxidants as the magic bullet you seek, do you consider the other potentially hundreds of variables (phytochemicals) involved, which are likely working synergistically together to get the desired result you are looking for?

The data has been quite clear on this issue for well over a decade. Giving individual, or a combination of, antioxidants as supplements do not provide the associated health or performance benefits they are correlated with. The individual antioxidants are not magical, may even be harmful, and must work synergistically with other chemicals found in produce and grains to obtain the desired benefits. As an example, on June 27, 2000, *ScienceDaily* published "Phytochemicals in Apples Are Found to Provide Anticancer and Antioxidant Benefits."[25] This report covered the Cornell University researcher Dr. Rui Hai Liu's work comparing the antioxidant potential of a whole apple vs. the actual vitamin C content of the apple. Dr. Liu's study demonstrated the following:

- "A combination of plant chemicals, such as flavonoids and polyphenols—collectively known as phytochemicals—found both within the flesh of apple and particularly in the skin—provide the fruit's antioxidant and anti-cancer benefits."
- "Scientists are interested in isolating single compounds—such as vitamin C, vitamin E and beta carotene—to see if they exhibit antioxidant or anti-cancer benefits. It turns out that none of those works alone to reduce cancer. It's the combination of flavonoids and polyphenols doing the work."
- "An antioxidant is one of many chemicals that reduce or prevent oxidation, thus preventing cell and tissue damage from free radicals in the body."
- "In this research, we have shown the importance of phytochemicals to human health," says Liu's collaborator, Chang Yong Lee, Cornell professor of food science at the university's New York State Agricultural Experiment Station in Geneva, New York. "Some of the phytochemicals are known to be anti-allergenic, some are anti-carcinogenic, anti-inflammatory, anti-viral, anti-proliferative. Now I have a reason to say an apple a day keeps the doctor away."
- "The researchers found that vitamin C in apples is only responsible for a small portion of the antioxidant activity. Instead, almost all this activity in apples is from phytochemicals. Indeed, previous studies have shown that a 500mg vitamin C pill might act as a pro-oxidant. The Cornell researchers found that eating 100 grams of fresh apple with skins provided the total antioxidant activity equal to 1,500 milligrams of vitamin C."

The Cornell University research involved looking at colon cancer cells that were treated with apple extracts. The biggest effect (i.e. inhibition of growth of these cells) was seen in the cells that were treated with extracts from both the skin as well as the fleshy part of the apple. The 100g apple used in the Cornell study is a very small apple. On your next trip to the grocery, weigh one small apple. Most weigh roughly a quarter pound, a little over 100 grams. An apple this size has approximately 8mg of vitamin C and the Cornell researchers found that the apple provided a total antioxidant activity equal to 1,500mg of vitamin C, as stated. This is a clear illustration of the positive synergistic effect of all plant chemicals vs. the supplementation of individual nutrients.

On March 1, 2005, the *Cornell Chronicle* published "Apples Could Help Reduce the Risk of Breast Cancer, Study Suggests."[26] Here they highlight additional comments of Dr. Liu regarding the food vs. supplement issue. Dr. Liu states, "Studies increasingly provide evidence that it is the additive and synergistic effects of the phytochemicals present in fruits and vegetables that are responsible for their potent antioxidant and anticancer activities." He also stated, "our findings suggest that consumers may gain more significant health benefits by eating more fruits and vegetables and whole grain foods than in consuming expensive dietary supplements, which do not contain the same array of balanced, complex components." And, most importantly, "he noted that the thousands of phytochemicals in foods vary in molecular size, polarity and solubility, which could affect how they are absorbed and distributed in different cells, tissues and organs." This is a major point.[27]

The same article quotes David R. Jacobs, professor in the Division of Epidemiology, School of Public Health, University of Minnesota:

> Dr. Liu is in the forefront of a group of investigators, including myself, who find extensive evidence that extremely important health aspects of food work through the combination of substances that make up that food, a concept we call "food synergy." Risk of many chronic diseases in modern life appears to be reduced by whole foods, but not by isolated large doses of selected food compounds.[27]

In another example, in June 2004, the *American Journal of Clinical Nutrition* published "The 6-a-Day Study: Effects of Fruit and Vegetables on Markers of Oxidative Stress and Antioxidative Defense in Healthy Nonsmokers."[28] This study reported on a 25-day study using 43 healthy male and female nonsmokers that demonstrated that the biomarkers (a substance found in the blood that can be used to evaluate the effectiveness of a treatment or supplement) of oxidative damage to protein and lipids (fats) were significantly lower in individuals consuming 600g of fruits and vegetables per day compared to individuals who took a pill containing the vitamins and minerals corresponding to those in 600g of fruit and vegetables.

In 1994, the late Dr. Victor Herbert made the following comments in the 1994 *American Journal of Clinical Nutrition* commentary "The Antioxidant Supplement Myth."[29] Dr. Herbert is recognized as an expert by the Federation of American Societies for Experimental Biology and is a past professor of medicine at Mt. Sinai School of Medicine, New York City. I am providing this older quote, as well as many others I have used in this chapter, to illustrate just how long antioxidant supplements have been frowned upon by good science practices yet still maintain a high volume of sales among consumers.

Dr. Herbert says:

- "The nutrient buzzword for 1994 is 'antioxidant.' Every supplement so labeled is seen as having only an upside and no downside. This is a myth. No supplement is a pure antioxidant."
- "At the November 1993 Food and Drug Administration Conference on Antioxidant Vitamins in Cancer and Cardiovascular Disease, there was essentially unanimous agreement that vitamins C, E, and B-carotene are mischaracterized when they are described solely as 'antioxidants' (fighters against harmful free radicals). What they are, in fact, is redox agents, antioxidant in some circumstances (often so in the physiological quantities found in food), and pro-oxidant (producing billions of harmful free radicals) in other circumstances (often so in the pharmacological quantities found in supplements)."
- "Large doses of vitamin E enhance immune activity and thus may promote progression of immune and autoimmune disease (i.e. asthma, food allergy, diabetes, rheumatoid arthritis, multiple sclerosis, and lupus)."
- "Vitamin C is especially dangerous in the presence of high body iron stores, which make vitamin C violently pro-oxidant. For genetic reasons, more than 10% of American Caucasians and perhaps as many as 30% of American blacks have high body iron, a condition called hemochromatosis. For consumer protection, every advertisement and label for vitamin C and/or iron supplements should warn: Do not take this product until your blood iron status has been determined."
- "We recently reported that in the presence of iron, not only does vitamin C appear to be worthless again cancer, but it increased lipoxidation of relatively harmless low-density-lipoprotein (LDL) cholesterol to coronary-artery-damaging oxidized LDL cholesterol."
- "Because vitamin C has the ability to release or mobilize stored iron in the body, taking heavy vitamin C supplementation could cause iron release into the blood beyond the iron binding capacity. The resulting free iron could have severe consequences to heart tissue from iron overload."

Dr. Herbert also pointed out that the vitamin C in a supplement would not have the same effect as the vitamin C found in food; for example, a glass of orange juice. He points out that food always contains a balance of the natural mixture of vitamin C biochemistry, both oxidized and reduced forms, whereas supplements do not. He describes this supplemental vitamin as "unbalanced biochemistry."

In January 2004, the *Nutrition Journal* published a review entitled "Iron Supplements: The Quick Fix With Long-Term Consequences," from the School of Pharmacy and Biomolecular Sciences, University of Brighton, UK.[30] In this review, the authors make several important points:

- Vitamin C has been shown to exhibit antioxidant effects at low doses, but conversely at high doses, it becomes a pro-oxidant.
- The high intake of iron or vitamin C alone warrants serious consideration. In tandem, this cocktail is potent. Uncontrolled interaction between vitamin C and iron salts (free or unbound iron) leads to oxidative stress.
- Supplementing iron with vitamin C exacerbates oxidative stress in the gastrointestinal tract, leading to ulceration in healthy individuals.
- Further studies need to be conducted to examine the detrimental effects of nutraceuticals, especially in chronic inflammatory conditions.
- High tissue concentrations of iron are associated with several pathologies, including some cancers, inflammation, diabetes, liver and heart disease.

The points Dr. Herbert and the *Nutrition Journal* article make are considerations for athletes and consumers, especially males. Iron losses from the body are small, and most of it is recycled from molecules which contain it. Females will lose iron each month through menstruation, and it is not uncommon for low-meat-eating or vegetarian women to be anemic. However, for males, the main and only real regulation of iron is through its absorption from the intestines. If iron absorption exceeds the very small excretion rates of it through greater absorption, the long-term health consequences can be very serious.

The typical American diet is already too high in protein, and most athletes consume even higher quantities due to the misunderstanding of just how little protein is in one pound of muscle tissue (only 22%, or 100g—see Chapter 6 for details). The resulting excess protein intake, as well as iron in tandem with large doses of supplemental vitamin C, as part of their "anti-oxidant" muscle recovery cocktail, may lead to excessive iron absorption and retention even without the genetic disorder hemochromatosis.

According to the Centers for Disease Control and Prevention (CDC), the hereditary disorder hemochromatosis is one of the most common genetic disorders. The CDC states that some of the disorders associated with this disease are arthritis, cirrhosis of the liver, diabetes, heart failure, and liver cancer.[31] Individuals who have been diagnosed with this condition should be extremely cautious about taking vitamin C supplements, due to vitamin C capacity to increase the absorption rate of iron.

But Antioxidants Are Natural!

This phrase is as worn out as an old pair of shoes, and far more senseless to still be using. Whether the compound is natural or synthetic is a moot point when it comes to its safety to ingest. It's the dosage, not the chemical itself. Yes, you are going to hear it again and again: *The Principle of Toxicology—the dose makes the poison*. Additionally, there is nothing "natural" about supplements, as Ritva Butrum, Ph.D., vice president of research at the American Institute of Cancer Research, stated in the press release "Magic Bullet Supplements Unlikely to Prevent Cancer" in 2000:[32]

> There is no such thing as a natural supplement. It is a contradiction of terms. The natural thing would be to get these substances in the combinations and amounts that occur in a healthy, balanced diet. There is nothing remotely natural about a supplement containing a single compound in amounts five, ten, or twenty times greater than anything found in nature.

Displacement of Stored Antioxidants

The last issue regarding the potential negative effects of supplemental antioxidants is related to the possible displacement of other stored antioxidants.

Researchers have known for many years that excessive intakes of the alpha form of vitamin E can displace other forms of the vitamin. In 1994, researchers at the Department of Molecular and Cell Biology, University of California at Berkeley, reported in the *American Journal of Clinical Nutrition* article "Human Adipose Alpha-Tocopherol and Gamma-Tocopherol Kinetics During and After 1 Year of Alpha-Tocopherol Supplementation,"[33] that the displacement of stored gamma-tocopherol with the alpha form (found in supplements) with as little as 250mg per day.

In 1997, researchers examined the effects of alpha-tocopherol supplements on the storage of the gamma form of vitamin E in skeletal muscle tissue. This was reported in *The Journal of Nutritional Biochemistry* report "Muscle Uptake of Vitamin E and Its Association With Muscle Fiber Type."[34] The subjects were given 800IU of alpha-tocopherol per day for 30 days. The researchers found that the alpha supplements resulted in a 300% increase in plasma levels of the alpha form and gamma-tocopherol decreased by 74% within 15 days of supplementation. Muscles biopsies taken before and after supplementation demonstrated a 53% increase in alpha-tocopherol and a 37% decrease of the gamma-tocopherol form.

An important point to remember regarding supplemental vitamin E displacing other forms of it is that several free radicals are not neutralized by alpha-tocopherol. They respond instead to gamma-tocopherol, which is found in food (specifically in nuts, grain, and soybeans), but not in most supplements. This data was reported in the April 1997 *Proceedings of the National Academy of Sciences* by

Stephen Christen, Ph.D., a biochemist and researcher at the University of California at Berkeley.[35] He noted that only gamma-tocopherol neutralizes peroxynitrite, a very destructive free radical found at inflammation sites, which is a common occurrence in athletes who routinely overwork muscle tissue and their tendons. In addition, gamma-tocopherol appears to be responsible for the neutralization of nitrogen oxide, a common air pollutant. Athletes who train in polluted regions of the country should keep this in mind when they supplement with large doses of alpha-tocopherol.

The *Federation of American Societies for Experimental Biology Online Journal* also reported on this potentially significant problem of supplemental alpha-tocopherol vitamin E displacing other biologically important forms. The 2003 report "Gamma-Tocopherol Inhibits Human Cancer Cell Cycle Progression and Cell Proliferation by Down-Regulation of Cyclins" stated that the gamma form of vitamin E demonstrated a more significant growth inhibition or anticancer activity on prostate, colorectal, and bone cancer cells than did the alpha form found in most supplements.[36] The authors speculated this was *unrelated to any antioxidant activity*. Remember, all nutrients are involved in more than one specific task.

It also appears that vitamin C supplements can displace other naturally produced antioxidants. In a November 1, 2002 study published in *Free Radical Biology and Medicine*, entitled "UVR-Induced Oxidative Stress in Human Skin in Vivo: Effects of Oral Vitamin C Supplementation,"[37] researchers examined the effects of giving participants 500mg/day of vitamin C for eight weeks prior to inducing ultraviolet light radiation (UVR) damage to the skin. The purpose was to assess if the vitamin C supplements would reduce the effect of the UVR. What the authors found was that the supplemental vitamin C not only had no effect of reducing the damage of UVR on the skin; it resulted in the reduction of several other antioxidants normally found and stored in skin cells.

In a 2001 review article in the *American Journal of Clinical Nutrition* entitled "Gamma-Tocopherol, the Major Form of Vitamin E in the U.S. Diet, Deserves More Attention," made the following points.[38] Again, the article further illustrates the foolishness of assuming the "more is better" mentality which supplement users embrace:

- Despite the fact that various forms of vitamin E have been identified, alpha-tocopherol is the only form that has been extensively studied and is present in most supplements.
- Gamma-tocopherol, being the major form of vitamin E in many plant seeds, is unique in many aspects. Compared with alpha-tocopherol, gamma-tocopherol is a slightly less potent antioxidant with regard to electron-donating propensity, but is superior in detoxifying various free radicals.
- We propose that although alpha-tocopherol is certainly a very important component of vitamin E, gamma-tocopherol may contribute significantly to human health in ways that have not yet been recognized.

- Because large doses of alpha-tocopherol are known to deplete plasma and tissue gamma-tocopherol, it is our opinion that this possibility should be considered and carefully evaluated.
- Recent evidence suggests that gamma-tocopherol has properties that may be important to human health and that are not shared by alpha-tocopherol.
- The qualities that distinguish gamma-tocopherol from alpha-tocopherol are likely a result of its distinct chemical reactivity, metabolism, and biological activity.

The Main Take-Home Messages

1. You need a balance of both oxidation and antioxidation reactions to be healthy. Free radicals are a normal part of cellular metabolism.
2. The supplementation of high dosages of the popular "antioxidants" might disrupt the normal balance of these compounds in the body, may displace other relevant antioxidants who have their own specific biological roles from storage, and likely cause more harm than good.
3. There are likely hundreds of naturally occurring antioxidants most of which we have very little understanding of. So, it is impossible for any supplement manufacturer to have a clue as to what really constitutes a "complete antioxidant formula," as many will attempt to do.
4. Clearly, there are no magic bullets or panaceas which can be bottled and swallowed that would replace wise food choices. However, the potential downside of antioxidants supplements are never mentioned in advertising.
5. If you still believe the supplement industry will provide you with accurate information in this area, consider the following from an article, "Antioxidants for Athletes!," posted on bodybuilding.com.[39] Here are a few points from the article:

   - "Most of the research has focused on Vitamin C and Vitamin E—and although the data is mixed, it is pretty convincing that these two powerful antioxidants, when taken in moderate doses, can help to prevent cell damage in response to free radicals." Just the opposite of what was presented in this chapter.
   - "With the beneficial effects of antioxidants demonstrated in athletes, researchers have begun to examine other antioxidants for their protective effects." Again, just the opposite of what was presented in this chapter.
   - "I would suggest 500–1000mg of Vitamin C and 400IU of Vitamin E is a good first step in antioxidant protection." A useless and pointless regimen.

Notes

1. The Truth about Human Aging. *Scientific American* (May 13, 2002). https://www.scientificamerican.com/article/the-truth-about-human-agi/
2. Ibid, p. 1.

3. https://www.scientificamerican.com/article/antioxidants/
4. M.T. Dutra, S. Alex, A.F. Silva, L.E. Brown, M. Bottaro, Antioxidant supplementation impairs changes in body composition induced by strength training in young women. *International Journal of Exercise Science* (March 1, 2019),Vol. 12, No. 2, pp. 287–96.
5. M.K. Ranchordas, D. Rogerson, H. Soltani, J.T. Costello. Do antioxidant supplements reduce muscle soreness after exercise? *British Journal of Sports Medicine-Blog* (July 27,2018). https://blogs.bmj.com/bjsm/2018/03/09/do-antioxidant-supplements-prevent-or-reduce-muscle-soreness-after-exercise/
6. M.T. Dutra, S. Alex, M.R. Mota, N.B. Sales, L.E. Brown, and M. Bottaro, Effect of strength training combined with antioxidant supplementation on muscular performance. *Journal of Applied Physiology, Nutrition, and Metabolism* (August 2018), Vol. 43, No. 8, pp. 775–81. https://www.ncbi.nlm.nih.gov/pubmed/29939770
7. Wiley, Beware of antioxidant supplements, warns scientific review. *ScienceDaily* (July 19, 2016). https://www.sciencedaily.com/releases/2016/07/160719094130.htm
8. Peter Tarr, Novel drug therapy kills pancreatic cancer cells by reducing levels of antioxidants: Strategy based on mimicking the suppression of antioxidant-promoting NRF2. *Cold Spring Harbor Laboratory Science Daily* (July 28, 2016). https://www.sciencedaily.com/releases/2016/07/160728143257.htm
9. T.L. Merry, and M. Ristow, Do antioxidant supplements interfere with skeletal muscle adaptation to exercise training? *Journal of Physiology* (2016),Vol. 594, pp. 5135–47.
10. Maria-Carmen Gomez-Cabrera, José Viña & Li Li Ji, Interplay of oxidants and antioxidants during exercise: implications for muscle health. *The Physician and Sportsmedicine* (2009),Vol. 37, No. 4, pp. 116–23.
11. Peter Tarr, How antioxidants can accelerate cancers, and why they don't protect against them. *Cold Spring Harbor Laboratory ScienceDaily* (July 10, 2014). www.sciencedaily.com/releases/2014/07/140710094434.htm
12. S.E. Hartmann and S.C. Forbes, Attenuated effects of exercise with an antioxidant supplement: too much of a good thing? *The Journal of Physiology* (2014),Vol. 592, No. 2, pp. 255–6. https://www.ncbi.nlm.nih.gov/pmc/articles/PMC3922488/
13. Mari Carmen Gomez-Cabrera, Michael Ristow, and Jose Vina, Antioxidant supplements in exercise: worse than useless? *American Journal of Physiology, Endocrinology and Metabolism* (2012),Vol. 302, No. 4, pp. E476–7.
14. Tina Peternelj and Jeff Coombes, Antioxidant supplementation during exercise training: beneficial or detrimental? *Sports Medicine* (2011),Vol. 41, pp.1043–69.
15. https://sciencebasedmedicine.org/antioxidants-and-exercise-more-harm-than-good/
16. M. Ristow, and S. Schmeisser, Extending life span by increasing oxidative stress, *Free Radical Biology and Medicine*, (July 15, 2011),Vol. 51, No. 2, pp. 327–36.
17. C.Y fanti, et al., Antioxidant supplementation does not alter endurance training adaptation, *Medicine & Science in Sports & Exercise* (July 2010),Vol. 4, No. 7, pp. 1388–95.
18. Jaouad Bouayed and Torsten Bohn, Exogenous antioxidants—double-edged swords in cellular redox state: health beneficial effects at physiologic doses versus deleterious effects at high doses. *Oxidative Medicine and Cellular Longevity* (2010),Vol. 3, pp. 228–37.
19. Cian McGinley, Amir Shafat, and Alan Donnelly, Does antioxidant vitamin supplementation protect against muscle damage? *Sports Medicine* (2009),Vol. 39, pp. 1011–32.
20. Mari Gomez-Cabrera, Elena Domenech, Marco Romagnoli, Alessandro Arduini, Consuelo Borras, Federico Pallardó, Juan Sastre, and Jose Viña, Oral administration of vitamin C decreases muscle mitochondrial biogenesis and hampers training-induced adaptations in endurance performance. *The American Journal of Clinical Nutrition* (2008), Vol. 87, pp. 142–9.
21. W.L Knez, D.G. Jenkins, and J.S. Coombes, Oxidative stress in half and full Ironman triathletes, *Medicine and Science in Sport and Exercise* (February 2007),Vol. 39, No. 2, pp. 283–8.
22. Ekhard E. Ziegler and L.J. Filer, Jr., eds., *Present Knowledge in Nutrition* (PKN), 7th ed. (Washington, DC: ILSI Press, 1996).

23. Maurice E. Shils, James A. Olson, and Moshe Shike, *Modern Nutrition in Health and Disease* (MNHD), 8th ed. (Philadelphia, PA: Lea & Febiger, 1994).
24. A. Catherine Ross, Benjamin Caballero, Robert Cousins, Katherine Tucker, and Thomas Ziegler, *Modern Nutrition in Health and Disease* (MNHD), 11th ed. (Wolters Kluwer Lippincott Williams & Wilkins).
25. Blaine Frielander, Phytochemicals in apples are found to provide anticancer and antioxidant benefits, Cornell researchers show. *Cornell University, ScienceDaily* (June 21, 2000). www.sciencedaily.com/releases/2000/06/000625232319.htm (accessed September 20, 2019).
26. Apples could help reduce the risk of breast cancer, study suggests. *Cornell Chronicle* (March 1, 2005). http://news.cornell.edu/stories/2005/03/apples-could-help-reduce-risk-breast-cancer-study-suggests
27. Ibid, p. 1.
28. Lars Dragsted, Anette Pedersen, Albin Hermetter, Samar Basu, Max Hansen, Gitte Ravn-Haren, Morten Kall, Vibeke Breinholt, Jacqueline Castenmiller, Jan Stagsted, Jette Jakobsen, Leif Skibsted, Salka Rasmussen, Steffen Loft, and Brittmarie Sandström, The 6-a-day study: Effects of fruit and vegetables on markers of oxidative stress and antioxidative defense in healthy nonsmokers. *The American Journal of Clinical Nutrition* (2004).Vol. 79, pp. 1060–72.
29. V. Herbert, The antioxidant supplement myth. *The American Journal of Clinical Nutrition* (August 1994),Vol. 60, No. 2, pp. 157–8.
30. Anna E.O. Fisher, and Declan P. Naughton, Iron supplements: The quick fix with long-term consequences. *Nutrition Journal* (January 2004),Vol. 3, No. 2.
31. https://www.cdc.gov/genomics/resources/diseases/hemochromatosis.htm
32. https://www.newswise.com/articles/magic-bullets-unlikely-to-prevent-cancer
33. G.J. Handelman, W.L. Epstein, J. Peerson, D. Spiegelman, L.J. Machlin, E.A. Dratz, Human adipose α-tocopherol and γ-tocopherol kinetics during and after 1 y of α-tocopherol supplementation, *The American Journal of Clinical Nutrition*, (May 1994), Vol. 59, No. 5, pp. 1025–32.
34. Mohsen Meydani, Roger A. Fielding, Joseph G. Cannon, Jeffrey B. Blumberg, and William J. Evans, Muscle uptake of vitamin E and its association with muscle fiber type. *The Journal of Nutritional Biochemistry* (1997),Vol. 8, No. 2, pp. 74–8.
35. Stephan Christen, Alan A. Woodall, Mark K. Shigenaga, Peter T. Southwell-Keely, Mark W. Duncan, and Bruce N. Ames, γ-Tocopherol traps mutagenic electrophiles such as NO_x and complements α-tocopherol: Physiological implications. *Proceedings of the National Academy of Sciences* (April 1997),Vol. 94, No. 7, pp. 3217–22.
36. Rene Gysin, Angelo Azzi, and Theresa Visarius, Gamma-tocopherol inhibits human cancer cell cycle progression and cell proliferation by down-regulation of cyclins. *FASEB Journal: Official Publication of the Federation of American Societies for Experimental Biology* (2003),Vol. 16, pp. 1952–4.
37. F. McArdle, L.E. Rhodes, R. Parslew, C.I.A. Jack, Peter Friedmann, and M.J. Jackson, UVR-induced oxidative stress in human skin in vivo: Effects of oral vitamin C supplementation. *Free Radical Biology & Medicine* (2002),Vol. 33, pp. 1355–62.
38. Q. Jiang, S. Christen, M.K. Shigenaga, and B.N. Ames, Gamma-tocopherol, the major form of vitamin E in the U.S. diet, deserves more attention, *American Journal of Clinical Nutrition* (December 2001),Vol. 74, No. 6, pp. 714–22.
39. www.bodybuilding.com/fun/berardi50.htm

9

SUPPLEMENTS

Are You Playing Russian Roulette With Your Health?

For the last 20 years, I have spoken three times a year at Kern High School District's coaches' clinic. The clinic is held each fall, winter, and spring for all new coaching staff. The clinic involves legal advice from a local attorney who makes coaches aware of the allowable do's and don'ts regarding what is considered acceptable behavior and actions coaches can take under a wide variety of circumstances. A local certified athletic trainer or physical therapist will provide the do's and don'ts regarding acceptable responses to injuries and the prevention of them. My role is to make the coaches aware of the do's and don'ts regarding nutrition and hydration issues, the potential serious liability issues associated with many of the supplement industry's products, and the deceptive marketing methods used by the supplement industry to exploit and fleece the many misunderstandings most consumers and athletes embrace. Further, my job is to illustrate how those misconceptions may negatively impact the athlete's or consumer's health, physical development, or performance. During the liability section of this lecture, I make it very clear as to why coaches should not, under any circumstances, recommend any over-the-counter product to any athlete accept for rehydration drinks such as Gatorade, Powerade, milk, and water, and of course food, due to the inherent liability with many supplemental products. This chapter will provide the reasoning for this opinion.

All active consumers and athletes, as well as those who advise them, should be aware of the following information which has come from the US Department of Justice, the US Food and Drug Administration (FDA), the United States Anti-Doping Agency (USADA), and peer-reviewed science journals. The following information is not exhaustive concerning this topic, but it should be enough to illustrate, as the title of this chapter declares, that for many supplements, those who ingest many of them are literally playing Russian roulette with their health.

One of the main illustrations of this chapter will be to highlight that the following information is a chronic problem with the supplement industry, and not some isolated problem with a few rogue suppliers over a brief period. Specifically, in 2006, when I published *Muscles Speed and Lies—What the Sport Supplement Industry Does Not Want Athletes or Consumers to Know*, I highlighted this issue with similar information which had been made public up to 2006. This book now illustrates that the supplement industry's modus operandi has not changed since 2006. For the past 13 years, it has remained defiant in policing itself to provide safe and effective products vs. a purely profit-driven industry with the motto "consumer beware—take at your own risk."

Dr. Pieter Cohen is an assistant professor at Harvard Medical School and an internist at Cambridge Health Alliance. In one of the educational videos, *Supplement 411*, provided later from the USADA web site, he states, "a lot of dangerous supplements are out there but we don't know which ones they are." Dr. Cohen consumer awareness work in this area, illustrating with his research the potential significant health problems associated with many of these products, has brought him into the crosshairs of the supplement industry, who would like to silence him as well as others who dare to point out the inherent liability with many products. To illustrate this point, STAT news service investigative report, "How a Supplement Maker Tried to Silence a Doctor," should be read.[1]

According to the report, a supplement maker sued Dr. Cohen for libel and slander after he pointed out in one of his published research papers that the products manufactured by the supplement maker "contained an illegal and potentially dangerous molecule, similar in structure to amphetamines." Dr. Cohen won the lawsuit, but the account of his case will illustrate the extent some in the supplement industry will go to in order to intimidate those who wish to expose the liability issue. Again, as stated previously, "consumer beware—take at your own risk."

Following, I provide two videos from USADA, two from the FDA, and the last from PBS *Frontline*. All will assist the consumer's and athlete's understanding of the deceptive practices of this industry and will provide educators an additional tool in the classroom to explore this issue.

- USADA Supplement 411 with Pieter Cohen M.D.: https://youtu.be/vIEOSQ_bDdE.
- USADA video using fictional product to teach how to decipher a supplement label: https://youtu.be/50QBwi11ncE.
- FDA: Warning on Body Building Products (Consumer Update): https://youtu.be/S9gXwOu2_sM.
- FDA video: "Stacking" Can Be Dangerous: www.youtube.com/watch?v=2zPhtpoAe40&feature=youtu.be&list=PLey4Qe-UxcxadPd4Ei1cVWKp-SJYYa1x6N.
- PBS *Frontline* broadcast on the safety of supplements: www.pbs.org/video/frontline-supplements-and-safety/.

• The transcript of the PBS *Frontline* broadcast can be found here: www.pbs. org/wgbh/frontline/film/supplements-and-safety/transcript/.

Instead of paraphrasing the many reports which illustrate why consumers should be very cautious in what they decide to consume to purportedly enhance their physique, physical performance, energy levels, strength, weight loss, etc., I am going to provide the information using the researcher's or organization's own words in order to enable the reader to fully grasp the extent of this problem. In this way, the reader should clearly see that this is not just my personal bias regarding this issue developed over the last 30 years, but a very broad professional opinion by those who understand this subject. I will provide reports from 2008 to 2019.

June 19, 2008

Journal of Mass Spectrometry

"Nutritional Supplements Cross-Contaminated and Faked With Doping Substances"[2]

In the Abstract:

Since 1999 several groups have analyzed nutritional supplements with mass spectrometric methods for contaminations and adulterations with doping substances.

These investigations showed that nutritional supplements contained prohibited stimulants as ephedrines, caffeine, methylenedioxymetamphetamie and sibutramine, which were not declared on the labels. An international study performed in 2001 and 2002 on 634 nutritional supplements that were purchased in 13 different countries showed that about 15% of the nonhormonal nutritional supplements were contaminated with anabolic-androgenic steroids (mainly prohormones). Since 2002, also products intentionally faked with high amounts of "classic" anabolic steroids such as metandienone, stanozolol, boldenone, dehydrochloromethyl-testosterone, oxandrolone etc. have been detected on the nutritional supplement market. These anabolic steroids were not declared on the labels either. The sources of these anabolic steroids are probably Chinese pharmaceutical companies, which sell bulk material of anabolic steroids. In 2005, vitamin C, multivitamin and magnesium tablets were confiscated, which contained cross-contaminations of stanozolol and metandienone. Since 2002 new "designer" steroids such as prostanozol, methasterone, androstatrienedione etc. have been offered on the nutritional supplement market. In the near future also cross-contaminations with these steroids are expected. Recently a nutritional supplement for weight loss was found to contain the β2-agonist clenbuterol. The application of such nutritional supplements is connected with a high risk of inadvertent doping cases and a health risk. For the detection of new 'designer' steroids in nutritional supplements, mass spectrometric strategies (GC/MS, LC/MS/MS) are presented.

May 27, 2013

Journal of the American Medical Association Internal Medicine

"The Frequency and Characteristics of Dietary Recalls in the United States"[3]

In the Results Section:

From January 1, 2004, through December 19, 2012, 465 drugs were subject to a class I recall in the United States. Just over one-half (237 [51%]) were classified as dietary supplements as opposed to pharmaceutical products. Most recalls occurred after 2008 (210 [89%]). Supplements marketed as sexual enhancement products (95 [40%]) were the most commonly recalled dietary supplement product, followed by bodybuilding (73 [31%]) and weight loss products (64 [27%]). Unapproved drug ingredients (237) accounted for all recalls.

In the Discussion Section:

Recalls of dietary supplements containing unapproved pharmaceutical ingredients are increasing. With over 150 million US residents consuming these products, the challenges posed by this growing and unregulated industry are enormous. To protect the health and safety of the public, increased efforts are needed to regulate this industry through more stringent enforcement and a standard of regulation similar to that for pharmaceuticals.

September 2013

Journal of Liver International

"Hepatotoxicity From Anabolic Androgenic Steroids Marketed as Dietary Supplements"[4]

We describe two Caucasian males (aged 25 and 45 years) with cholestatic hepatitis following ingestion of the dietary supplement Mass-Drol ("Celtic

Dragon") containing androgenic anabolic steroids. Androgenic anabolic steroids marketed as dietary supplements continue to cause hepatotoxicity in the UK.

October 14, 2013

Journal of Drug Testing and Analysis

"Muscles and Meth: Drug Analog Identified in 'Craze' Workout Supplement"[5]

An international team of scientists have identified potentially dangerous amounts of methamphetamine analog in the workout supplement Craze, a product widely sold across the U.S. and online. The study, published in Drug Testing and Analysis, was prompted by a spate of failed athletic drug tests. The results reveal the presence of methamphetamine analog N,α-DEPEA, which has not been safely tested for human consumption, in three samples.

In recent years banned and untested drugs have been found in hundreds of dietary supplements. We began our study of Craze after several athletes failed urine drug tests because of a new methamphetamine analog," said lead author Dr. Pieter Cohen, of Harvard Medical School, U.S.A.

A workout supplement marketed as a 'performance fuel,' Craze is manufactured by Driven Sports, Inc., and is sold in stores across the United States and internationally via body supplement websites.

The supplement is labeled as containing the compound N,N-diethylphenylethylamine (N,N-DEPEA), claiming it is be derived from endangered dendrobium orchids. However, while there is no proof that this compound is found within orchids, it is also structurally similar to the methamphetamine analog N,α-diethylphenylethylamine (N,α-DEPEA), a banned substance.

The team analyzed three samples of Craze for traces of N,α-DEPEA. The first sample was brought from a mainstream retailer in the U.S., while the second and third samples were ordered from online retailers in the U.S. and Holland.

The team used ultra-high performance liquid chromatography to detect the presence of N,α-DEPEA. The first two samples were analyzed by NSF International, while the third was tested at the Netherland's National Institute for Public Health. The findings were independently corroborated by the Korean Forensic Service, which confirmed the presence of N,α-DEPEA in two further samples of Craze in a parallel investigation.

"We identified a potentially dangerous designer drug in three separate samples of this widely available dietary supplement," said Cohen. "The tests revealed quantities of N,α-DEPEA of over 20mg per serving, which strongly suggests that this is not an accidental contamination from the manufacturing process."

As a structural analog of methamphetamine, N,α-DEPEA, may have stimulant and addictive qualities; however, it has never been studied in humans and its adverse effects remain unknown.

The product labeling claims that Craze contains several organic compounds, known as phenylethylamines. However, phenylethylamines are a very broad category of chemicals which range from harmless compounds found in chocolate to synthetically produced illegal drugs.

"The phenylethylamine we identified in Craze, N,α-DEPEA, is not listed on the labeling and it has not been previously identified as a derivative of dendrobium orchids," said Cohen.

"If these findings are confirmed by regulatory authorities, the FDA (The U.S. Food and Drug Administration) must take action to warn consumers and to remove supplements containing N,α-DEPEA from sale," concluded Cohen. "Our fear is that the federal shutdown may delay this, resulting in potentially dangerous supplements remaining widely available."

May 27, 2013

Journal of the American Medical Association, Internal Medicine

"How Can We Know If Supplements Are Safe If We Do Not Know What Is in Them?"[6]

Comment on "The Frequency and Characteristics of Dietary Supplement Recalls in the United States"

Americans spend over $20 billion annually on dietary supplements. Although supplements are regulated by the US Food and Drug Administration (FDA) under the Dietary Supplement Health and Education Act, there is no requirement for supplement manufacturers to demonstrate efficacy or safety of their products prior to marketing them. However, companies may not include unapproved ingredients. It turns out that even this minimal requirement is not fulfilled. Harel et al identified 237 dietary supplements that were recalled by the FDA owing to inclusion of unapproved drug ingredients. Given the limited regulation of these products, it is likely that the number of recalls grossly underestimates the number of products on sale with unapproved ingredients. Dietary supplements should be treated with the same rigor as pharmaceutical drugs and with the same goal: to protect consumer health.

October 2014

Harvard Public Health Review, Volume 2

"How America's Flawed Supplement Law Creates the Mirage of Weight Loss Cures"[7]

By Pieter Cohen, M.D.

Dr. Mehmet Oz, "America's Doctor," recently testified before Congress on why he was promoting supplements containing Garcinia cambogia, raspberry ketones, green coffee extract and other unproven ingredients as weight loss miracles. He acknowledged that "sometimes they don't have the scientific muster to pass as fact". In fact, there is no legal over-the-counter botanical supplement that has demonstrated clinical efficacy as a diet pill. [The only herbal treatment that can lead to modest weight loss is ephedra combined with caffeine, but this cocktail can also cause strokes, heart attacks and sudden death; hence, ephedra was banned in 2004.]

Promoting worthless supplements as weight loss miracles is irresponsible for any physician, especially those with the public influence such as Dr. Oz, but the problem with weight loss supplements is much larger than even "America's Doctor".

Any supplement sold in the United States can be legally promoted for weight loss without demonstrating evidence of its efficacy or safety through human subject studies. The law governing supplements, the Dietary Supplement Health and Education Act of 1994 (DSHEA), shields supplements from scientific scrutiny while permitting their sale as weight loss pills. All supplements are presumed safe until the US Food and Drug Administration (FDA) proves otherwise. Unencumbered by the need to study efficacy or safety, or even to disclose adverse effects, the supplement industry has flourished. In 2012 Americans spent $32 billion on more than 85,000 varieties of pills, powders and potions labeled as dietary supplements. DSHEA has enabled thriving trade in ineffective and potentially harmful weight loss supplements at GNC, Vitamin Shoppe, chain pharmacies, supermarkets and countless other stores throughout America.

The herbal ingredients in weight loss supplements may pose health risks. Consider the example of yohimbe, an African shrub, found in many weight loss supplements. It can cause elevated blood pressure or panic attacks. In addition to adverse effects from the herbal ingredients listed on the label, weight loss supplements have frequently been found to contain illegal drugs. DSHEA creates perverse incentives for unscrupulous manufacturers to out-compete legitimate companies by adding undeclared, illegal ingredients including prescription medications, banned drugs and even entirely novel chemical compounds. Although the FDA has only tested a very small proportion of all supplements, the agency has already found hundreds of weight loss supplements tainted with banned drugs. Some of these illegal drugs may lead to short-term weight loss, but they often carry the risk of serious health consequences.

The combination of old and new ingredients in weight loss supplements can pose serious health risks. Last summer, a top liver transplant surgeon in Hawaii became increasingly alarmed as she cared for an onslaught of new patients with unexplained liver failure. Epidemiologists from the Centers for Disease Control and Prevention (CDC) identified the cause of liver failure to be a weight loss supplement containing a cocktail of old and new ingredients called OxyElite Pro. The CDC investigation eventually found nearly one hundred people in sixteen states who had developed hepatitis from this one supplement. Forty-seven required hospitalization and three needed new livers. One mother of seven died. Unfortunately, OxyElite Pro is not alone in posing life-threatening risks. A national team of researchers recently revealed that the rate of liver failure from supplements has increased 185% over the past decade. Twenty percent of severe liver damage from drugs is now estimated to be caused by supplements.

Liver damage is not the only life-threatening risk from slimming supplements. New pharmaceutical stimulants never before tested in humans, including versions of amphetamine and methamphetamine, are currently found in a variety of weight loss supplements. One such example is 1,3-dimethylamylamine (DMAA), which has never been approved as an oral drug but is available over-the-counter in dozens of supplements. DMAA is currently being investigated as a cause of strokes, heart disease and sudden death.

The FDA is tasked with finding and removing these dangerous supplements but only after they are available on store shelves. Currently, the FDA does not have an effective system to detect supplements that pose serious health risks. Instead, the FDA relies on consumers and physicians to voluntarily submit reports of harm from supplements at www.safetyreporting. hhs.gov. However, few consumers or doctors contact the agency or even know that the FDA regulates supplements. The US Government Accountability Office estimates that a small fraction of the estimated 50,000 adverse reactions each year from supplements are reported to the FDA. The lack of reporting, along with the poor quality of the information received in these few reports make it nearly impossible for the FDA to find and remove dangerous supplements.

Should all supplements require FDA approval? If a supplement does not claim to improve one's health (in the parlance of DSHEA, advertised with "structure/function" claims) then there is no reason, in my opinion, to require FDA evaluation of efficacy. However, if supplements are promoted as if they will improve health or performance—for example, stimulate weight loss, decrease blood sugar, enhance athletic performance—then rigorous evidence of efficacy and safety should be vetted by the FDA before these supplements are permitted to be sold on store shelves. DSHEA, the law that has enabled countless dangerous weight loss pills to reach store shelves, needs to be reformed. To prevent serious harm, it's time that rigorous evidence of efficacy and safety be required before any pills, powders or potions may be sold as over-the-counter weight loss supplements.

July 12, 2014

Journal of Hepatology

"Liver Injury From Herbals and Dietary Supplements in the U.S. Drug-Induced Liver Injury Network"[8]

In the Abstract:

The Drug-Induced Liver Injury Network (DILIN) studies hepatotoxicity caused by conventional medications as well as herbals and dietary supplements (HDS). To characterize hepatotoxicity and its outcomes from HDS versus medications, patients with hepatotoxicity attributed to medications or HDS were enrolled prospectively between 2004 and 2013. The study took place among eight U.S. referral centers that are part of the DILIN. Consecutive patients with liver injury referred to a DILIN center were eligible. The final sample comprised 130 (15.5%) of all subjects enrolled (839) who were judged to have experienced liver injury caused by HDS.

In the Discussion Section:

Contrary to widespread belief, this study demonstrates that HDS products are not always safe. Indeed, our data suggest that, relative to conventional medication-induced hepatotoxicity, liver injury from HDS not only occurs, but also may be increasing in frequency over time in the populations surrounding the DILIN centers and, probably, in the United States as a whole. The study also shows that bodybuilding HDS are the most commonly implicated class of products. Most important, we found that nonbodybuilding HDS can cause liver injury that is more severe than conventional medications, as reflected in a higher transplantation rate.

June 17, 2014

The Food and Drug Administration

"Public Notification: Sport Burner Contains Hidden Drug Ingredient"[9]

Is advising consumers not to purchase or use Sport Burner, a product promoted and sold for weight loss on various websites and possibly in some retail stores.

FDA laboratory analysis confirmed that Sport Burner contains fluoxetine. Fluoxetine is an FDA approved drug in a class of drugs called selective serotonin reuptake inhibitors (SSRIs) used for treating depression, bulimia, obsessive-compulsive disorder (OCD), panic disorder, and premenstrual dysphonic disorder (PMDD).

Uses of SSRIs have been associated with serious side effects including suicidal thinking, abnormal bleeding, and seizures. In patients on other medications for common conditions (aspirin, ibuprofen, or other drugs for depression, anxiety, bipolar illness, blood clots, chemotherapy, heart conditions, and psychosis), ventricular arrhythmia or sudden death can occur.

April 9, 2014

New England Journal of Medicine

Perspective: "Hazards of Hindsight—Monitoring the Safety of Nutritional Supplements"[10]

Epidemiologists at the Centers for Disease Control and Prevention (CDC) recently confirmed what an astute liver-transplant surgeon in Honolulu already suspected: OxyElite Pro, a popular over-the-counter supplement, was responsible for a cluster of cases of severe hepatitis and liver failure. Although patients began to develop severe hepatitis in May 2013, the Food and Drug Administration (FDA), whose job it is to remove dangerous supplements from store shelves, did not learn of the cases until mid-September, 4 months later. By February 2014, the CDC had linked 97 cases, resulting in 47 hospitalizations, three liver transplantations, and one death, to OxyElite Pro. This dietary supplement was recalled, but nothing has been done to prevent another supplement from causing organ failure or death. Nor have any changes been made to improve the FDA's ability to detect dangerous supplements.

The FDA's delayed response—with its life-threatening consequences—is attributable to our woefully inadequate system for monitoring supplement safety. Americans spend more than $32 billion a year on more than 85,000 different combinations of vitamins, minerals, botanicals, amino acids, probiotics, and other supplement ingredients. Unlike prescription medications, supplements do not require premarketing approval before they reach store shelves. Under the Dietary Supplement Health and Education Act of 1994, anything labeled as a dietary supplement is assumed to be safe until proven otherwise. The FDA is charged with the unenviable task of identifying and removing dangerous supplements only after they have caused harm.

And the agency has its work cut out for it: potentially dangerous supplements are widely available. More than 500 supplements have already been found to be adulterated with pharmaceuticals or pharmaceutical analogues, including new stimulants, novel anabolic steroids, unapproved antidepressants, banned weight-loss medications, and untested sildenafil analogues. In 2013 alone, researchers discovered two new stimulants in widely marketed supplements. My colleagues and I identified a new analogue of methamphetamine, N,α-diethyl-phenylethylamine (N,α-DEPEA), in a popular sports supplement. FDA scientists discovered another stimulant, β-methylphenethylamine (β-MePEA)—a novel analogue of amphetamine—in nine supplements. N,α-DEPEA and β-MePEA have never been studied in humans, and their adverse effects are entirely unknown; yet they are sold as "natural" products without having undergone any premarketing testing for safety. (Although supplements containing N,α-DEPEA were voluntarily withdrawn from the market, supplements containing β-MePEA remain widely available).

October 15, 2015

New England Journal of Medicine

"Emergency Department Visits for Adverse Events Related to Dietary Supplements"[11]

We used nationally representative surveillance data from 63 emergency departments obtained from 2004 through 2013 to describe visits to U.S. emergency departments because of adverse events related to dietary supplements.

Symptoms Section:

Cardiac symptoms (palpitations, chest pain, or tachycardia) were the most common symptoms associated with weight-loss products (in 42.9% of patients) and energy products (in 46.0% of patients). Weight-loss or energy products were implicated in 71.8% of all emergency department visits for supplement-related adverse events involving palpitations, chest pain, or tachycardia. Most of the visits for cardiac symptoms involved persons 20 to 34 years of age. Cardiac symptoms were also commonly documented in emergency department visits attributed to bodybuilding products and sexual-enhancement products.

Discussion Section:

Limitations of our analysis should be noted. The number of emergency department visits attributed to supplement-related adverse events that we identified is probably an underestimation, since supplement use is underreported by patients, and physicians may not identify adverse events associated with supplements as often as they do those associated with pharmaceuticals. Physicians also may have more limited knowledge of interactions between prescription drugs and dietary supplements than they do interactions between prescription drugs. In addition, we did not collect data on emergency department visits associated with products that are generally considered to be foods or drinks by consumers but that may be considered to be dietary supplements under the Dietary Supplement Health and Education Act (e.g., energy drinks). However, it is also possible that emergency department physicians may incorrectly attribute certain symptoms to supplements, which could lead to overestimation.

Conclusions Section:

An estimated 23,000 emergency department visits in the United States every year are attributed to adverse events related to dietary supplements. Such visits commonly involve cardiovascular manifestations from weight-loss or energy products among young adults and swallowing problems, often associated with micronutrients, among older adults.

November 2, 2015

Journal of Drug Testing and Analysis

"Adulterated Dietary Supplements Threaten the Health and Sporting Career of Up-and-Coming Young Athletes"[12]

The authors provide the perspective of the United States Anti-Doping Agency on the actions youth sport organizations should take to protect the health of developing young athletes.

The anti-doping world is all too familiar with "sports" or "performance enhancing" dietary supplements (i.e., products marketed for body building, weight-loss, pre-workout/energy) that are adulterated with stimulants, anabolic agents, and pharmaceuticals, either by accidental contamination or by deliberate spiking. But the larger community surrounding developing young athletes is often not aware of the lurking health and anti-doping risks such products pose. Since most young athletes are not drug tested, they, along with the sports administrators, coaches, and parents, are often not motivated to read or are not presented with educational materials alerting them to the possibility of the adulteration of dietary supplements. They also assume that dietary supplements are safe. However, this simply isn't the case when it comes to sports supplements. Several reviews on the contents of such supplements have estimated a contamination rate with stimulants and/or anabolic-androgenic steroids of 14–18% and higher. This problem is only going to get worse as the global sports supplement portion of the industry grows and is estimated to reach $12 billion annually by 2020.

Most governments, including the US government, do not have effective regulations in place to protect consumers. A good example to illustrate this problem is the case of methylhexaneamine, an amphetamine-like stimulant first detected in 31 anti-doping samples in 2009 and subsequently linked to severe adverse events among athletes, military, and the general public. Despite multiple Food and Drug Administration (FDA) enforcement actions in 2013 and every year since, methylhexaneamine is still present in dietary supplements and causing serious harm. In another study, researchers examined 27 brands of supplements on sale after FDA recalls and found that 2 out of 3 were still adulterated with banned drugs. Despite attempts to introduce new legislation or tighten existing ones, the number of adulterated dietary supplements is ever present. The US government currently does not have effective tools to remove dangerous products from the market and therefore it is critical that coaches, parents, and others involved in administering youth sport organizations receive education about managing the risks posed by adulterated dietary supplements.

The true health consequences caused by adulterated supplements is unknown because adverse events are notoriously underreported. Many healthcare providers do not know how or where to file an adverse event report, and parents, coaches, and other youth sport administrators may not even notice the side-effects in their athletes, let alone recognize that they should report them. Furthermore, some side-effects are not acute events, but instead appear as delayed, chronic health problems. As an example, the long-term use of body building supplements has recently been linked to testicular cancer. Coaches and parents and the broader sport organization community cannot just rely on the sudden appearance of strange symptoms in their athletes to identify dangerous supplements.

Sport or performance-enhancing supplements pose risks that are unique to the growing athlete. The developing endocrine and nervous systems are sensitive to the presence of hormones, anabolic-androgenic steroids, and stimulants. For example, the endogenous production of testosterone can be suppressed for months following even a low dose of anabolic-androgenic steroids, and cardiovascular risks and mood disturbance are common side-effects for adolescents. Anabolic agents can also cause premature epiphyseal closure (which is permanent), brain remodeling, and an increased risk of maladaptive behaviors and neurological disorders. For stimulants, increases in blood pressure, loss of appetite, emotional instability, nervousness, jitteriness, and social withdrawal are common side-effects in youth, and prolonged exposure to stimulants can also negatively affect growth. To make matters worse, the stimulants currently popular in dietary supplements are untested in humans so their effects on the growing body and mind are completely undocumented. Pharmaceuticals, which are found surprisingly often in dietary supplements, can also have different toxicological effects on children than they do in adults, sometimes making their effects more severe, or at the very least unpredictable. Even very young children are at risk. A retrospective study of in children under 6 reported a 274% increase in the number of poisonings due to pharmaceuticals present in dietary supplements from 2000 to 2010. The substances commonly found in adulterated sports supplements pose serious risks to the health of young athletes but may not immediately induce obvious side-effects.

Even caffeine has serious and well-documented side-effects at high doses such as inducing transient ischemia and causing sudden death in otherwise healthy individuals. While many adolescents can handle moderate caffeine without serious consequences, it is often hard to identify the quantity of caffeine and other stimulants in dietary supplements or energy drinks. The American Academy of Pediatrics does not recommend caffeine for adolescents. Caffeine is a very popular ingredient in sports supplements, and it can be listed in obscure ways on the label, if it is listed at all. Caffeine should not be ignored in any conversation about dietary supplement safety for young athletes.

Exposure to adulterated sports supplements by adolescents and young adults may be high because the dietary supplements in question are often very popular. Among high school students, the level of supplement use can be as high as 74% and is correlated with the level of sport participation. Many young athletes report using the types of products that are at a higher risk of being adulterated, namely supplements to build muscle, lose weight, and to improve athletic performance. Adolescents report obtaining information about dietary supplements from coaches, friends, family members, teammates, the Internet, and other media. Fitness magazines often have high teen readership and contain many advertisements for dietary supplements that are false or misleading but are appealing to teens. The US Anti-Doping Agency (USADA) report—What Sport Means in America —also confirms that coaches are key in the life of a young athlete. It is essential that coaches and all of the other individuals surrounding the developing young athlete are well-informed on the dangers of sport supplements so that they can intervene if they learn about the use of such products by their athletes.

The stakes are high for all of us to protect the health of young athletes and to protect clean sport. Between the USADA Supplement 411 High Risk List and the FDA Health Fraud page, plus a multitude of reports in the scientific literature, there are more than a thousand products identified to contain stimulants, anabolic-androgenic steroids, or pharmaceuticals. While USADA will continue to support changes in legislation that would stop the sale of illegal and dangerous products under the guise of dietary supplements, we must act now to protect our stakeholders.

It is essential that coaches, parents, and all of the other support personnel involved in the administration of youth sports receive high-quality education about reducing the risks from the use of dietary supplements, especially for products marketed for body-building, weight-loss, or energy/ pre-workout. Educational efforts should include references to resources that list known adulterated or otherwise risky products such as the Supplement 411 High Risk List and the FDA Health Fraud page, and references to the importance of third-party certification. Youth sport administrators, including those administering sport programs in junior high and high schools, should have firm policies in place to limit access to such products, and should ensure that coaches and team trainers are aware of these policies.

This is a global problem and it is not going away anytime soon. Until governments around the world improve the effectiveness of enforcement tools to stem the tide of adulterated supplements, it is the responsibility of anti-doping agencies, and all sport organizations to protect clean athletes and clean sport by ensuring athletes do not use adulterated dietary supplements.

November 17, 2015

U.S. Department of Justice Office of Public Affairs

"Justice Department and Federal Partners Announce Enforcement Actions of Dietary Supplement Cases"[13]

"Criminal Charges Brought against Bestselling Supplement Manufacturer"

As part of a nationwide sweep, the Department of Justice and its federal partners have pursued civil and criminal cases against more than 100 makers and marketers of dietary supplements. The actions discussed today resulted from a year-long effort, beginning in November 2014, to focus enforcement resources in an area of the dietary supplement market that is causing increasing concern among health officials nationwide. In each case, the department or one of its federal partners allege the sale of supplements that contain ingredients other than those listed on the product label or the sale of products that make health or disease treatment claims that are unsupported by adequate scientific evidence.

Among the cases announced today is a criminal case charging USPlabs LLC and several of its corporate officers. USPlabs was known for its widely popular workout and weight loss supplements, which it sold under names such as Jack3d and OxyElite Pro.

The sweep includes federal court cases in 18 states and was announced today by Principal Deputy Assistant Attorney General Benjamin C. Mizer, head of the Justice Department's Civil Division; Deputy Commissioner for Global Regulatory Operations and Policy Howard Sklamberg J.D. of the Food and Drug Administration (FDA); Acting Deputy Director J. Reilly Dolan of the Federal Trade Commission (FTC)'s Bureau of Consumer Protection; Acting Deputy Chief Inspector Gary Barksdale of the U.S. Postal Inspection Service (USPIS); and Chief Richard Weber of the Internal Revenue Service's (IRS) Criminal Investigation (CI) Division. The Department of Defense (DoD) and the U.S. Anti-Doping Agency (USADA) are also participating in the sweep to unveil new tools to increase awareness of the risks unlawful dietary supplements pose to consumers and, in particular, to assist service members targeted by illegitimate athletic performance supplements.

"The Justice Department and its federal partners have joined forces to bringing to justice companies and individuals who profit from products that threaten consumer health," said Principal Deputy Assistant Attorney General Mizer. "The USPlabs case and others brought as part of this sweep illustrate alarming practices the department found—practices that must be brought to the public's attention, so consumers know the serious health risks of untested products."

During the period of the sweep, 117 individuals and entities were pursued through criminal and civil enforcement actions. Of these, 89 were the subject of cases filed since November 2014.

In the Criminal Matters Section:

The indictment alleges that USPlabs engaged in a conspiracy to import ingredients from China using false certificates of analysis and false labeling and then lied about the source and nature of those ingredients after it put them in its products. According to the indictment, USplabs told some of its retailers and wholesalers that it used natural plant extracts in products called Jack3d and OxyElite Pro, when in fact it was using a synthetic stimulant manufactured in a Chinese chemical factory.

The indictment also alleges that the defendants sold some of their products without determining whether they would be safe to use. In fact, as the indictment notes, the defendants knew of studies that linked the products to liver toxicity.

The indictment also alleges that in October 2013, USplabs and its principals told the FDA that it would stop distribution of OxyElite Pro after the product had been implicated in an outbreak of liver injuries. The indictment alleges that, despite this promise, USplabs engaged in a surreptitious, all-hands-on-deck effort to sell as much OxyElite Pro as it could as quickly as possible. It was sold at dietary supplement stores across the nation.

The charges and allegations in the indictments are merely accusations, and the defendants are presumed innocent unless and until proven guilty.

June 20, 2017

FDA Report

"Caution: Bodybuilding Products Can Be Risky"[14]

Your buddy at the gym can't say enough about the bodybuilding products he's been taking to help build muscle mass and strength. You wonder, are they all safe to use?

According to CDR Mark S. Miller, Pharm. D., a regulatory review officer at the U.S. Food and Drug Administration (FDA), bodybuilding products that contain steroids or steroid-like substances are associated with potentially serious health risks, including liver injury. "Some of the liver injuries were life-threatening," CDR Miller says.

CDR Miller was the lead reviewer assessing hundreds of adverse event reports made to the FDA from July 2009 through December 2016. Thirty-five reports showed evidence of serious liver injury.

In addition to liver injury, anabolic steroids have been associated with serious reactions such as severe acne, hair loss, altered mood, irritability, increased aggression, and depression. They have also been associated with life-threatening reactions such as kidney damage, heart attack, stroke, pulmonary embolism (blood clots in the lungs), and deep vein thrombosis (blood clots that occur in veins deep in the body).

These bodybuilding products are promoted as hormone products and/or as alternatives to anabolic steroids for increasing muscle mass and strength. Many of these products make claims about the ability of the active ingredients to enhance or diminish androgen, estrogen, or progestin-like effects in the body, but actually contain anabolic steroids or steroid-like substances, synthetic hormones related to the male hormone testosterone.

Cara Welch, Ph.D., a senior advisor in FDA's Office of Dietary Supplement Programs, explains that many of these bodybuilding products sold online as well as in retail stores, are labeled as "dietary supplements." "In fact," she says, "many of these products are not dietary supplements at all; they are illegally marketed, unapproved new drugs." FDA has not reviewed these products for safety, effectiveness, or quality before these companies began marketing.

CDR Jason Humbert, a regulatory operations officer in FDA's Office of Regulatory Affairs, says that potentially harmful, sometimes hidden ingredients in products promoted for body building continue to be a concern. "The companies making these products are breaking the law by exploiting an easily accessible marketplace to get these products to consumers," he says. "In the end, it's consumers who are put in harm's way by taking dangerous ingredients from products promoted as having miraculous results or making empty promises, and who may not understand the risks."

Some who use bodybuilding products engage in "stacking," using multiple products (including stimulants or products providing false assurances of liver protection) to enhance results or "gains." These combinations may put consumers at greater risk for serious and life-threatening reactions. Watch video here: www.youtube.com/watch?v=2zPhtpoAe40&feature=youtu. be&list=PLey4Qe-UxcxadPd4Ei1cVWKpSJYYa1x6N

October 31, 2017

The Food and Drug Administration

"FDA Warns Against Using SARMs in Body-Building Products"[15]

"We are extremely concerned about unscrupulous companies marketing body-building products with potentially dangerous ingredients. Body-building products that contain selective androgen receptor modulators, or SARMs, have not been approved by the FDA and are associated with serious safety concerns, including potential to increase the risk of heart attack or stroke and life-threatening reactions like liver damage," said Donald D. Ashley, J.D., director of the Office of Compliance in the FDA's

Center for Drug Evaluation and Research. "We will continue to take action against companies marketing these products to protect the public health."

The U.S. Food and Drug Administration recently issued warning letters to Infantry Labs, LLC, IronMagLabs and Panther Sports Nutrition for distributing products that contain SARMs. Although the products identified in the warning letters are marketed and labeled as dietary supplements, they are not dietary supplements. The products are unapproved drugs that have not been reviewed by the FDA for safety and effectiveness.

Life threatening reactions, including liver toxicity, have occurred in people taking products containing SARMs. SARMs also have the potential to increase the risk of heart attack and stroke, and the long-term effects on the body are unknown. Consumers should stop using these body-building products immediately and consult a health care professional if they are experiencing any adverse reactions that may be associated with their use.

June 20, 2017

Food and Drug Administration

"FDA Issues Warning About Body-Building Products Labeled to Contain Steroid and Steroid-Like Substances"[16]

The U.S. Food and Drug Administration today posted warning letters to Flex Fitness/Big Dan's Fitness, AndroPharm, and Hardcore Formulations for illegally marketed products labeled to contain steroid and steroid-like substances and promoted to increase muscle mass and strength.

Although the products that are the subject of these warning letters are marketed as dietary supplements, they are not dietary supplements. Instead, these products are unapproved drugs that FDA has not reviewed for safety and effectiveness.

FDA recommends consumers immediately stop using over-the-counter body-building products labeled or promoted to contain steroid and steroid-like substances due to the risk of serious liver injury and other adverse health consequences including kidney injury, increased risk of heart attack and stroke, and shrinkage of the testes and male infertility.

The agency reviewed more than seven years of adverse event reports and found 35 patients who suffered liver injuries, many requiring hospitalization, that were associated with these types of products.

November 3, 2017

Clinical Toxicology

"Four Experimental Stimulants Found in Sports and Weight Loss Supplements: 2-Amino-6-Methylheptane (Octodrine), 1,4-Dimethylamylamine (1,4-DMAA), 1,3-Dimethylamylamine (1,3-DMAA) and 1,3-Dimethylbutylamine (1,3-DMBA)"[17]

In the Introduction Section:

Epidemiologists at the Centers for Disease Control and Prevention (CDC) reported in 2015 that dietary supplements are responsible for tens of thousands of emergency department visits and thousands of hospitalizations each year in the United States (US). The CDC investigators found that weight loss and sports supplements accounted for a disproportionate number of emergency department visits, but why these particular categories of supplements pose greater risks to consumers is not well understood. One possibility is that botanical stimulants contained in such products, including caffeine and yohimbine, might contribute to these adverse events. Another possibility is that experimental, synthetic stimulants found in many brands of weight loss and sports supplements may contribute to the increased health risks of these products.

Experimental stimulants, including analogs of amphetamine, methamphetamine and ephedrine, have been found in hundreds of brands of weight loss and sports supplements. Until recently, the most common synthetic stimulant in supplements was 1,3-dimethylamylamine (1,3-DMAA). However, beginning in 2012 the US Food and Drug Administration (FDA) took enforcement actions to remove 1,3-DMAA from supplements due to a lack of evidence supporting it as a legal dietary ingredient and safety concerns. Shortly thereafter, 1,3-dimethylbutylamine (1,3-DMBA), an analog of 1,3-DMAA, was discovered in supplements. In 2015, the FDA took enforcement actions to remove 1,3-DMBA from supplements because the conditions for it to be lawfully marketed had not been met. Whether other analogs of 1,3-DMAA have been introduced as novel ingredients in sports and weight loss supplements since FDA took action to remove 1,3-DMAA and 1,3-DMBA from supplements remains unknown.

We designed the current study to determine if a new stimulant structurally similar to 1,3-DMAA was present as an ingredient in dietary supplements sold in the United States. For the purposes of this study, we will refer to banned stimulants as those ingredients for which the FDA had taken enforcement action to remove from dietary supplements prior to August 2016 (when the samples were purchased) and use experimental stimulants to refer to banned stimulants as well as those ingredients for which the FDA had not taken enforcement action to remove from dietary supplements prior to August 2016.

Conclusions Section:

In a study of dietary supplements, we found two stimulants which the FDA, the agency responsible for the regulation of supplements in the US, has attempted to remove from supplements, 1,3-DMAA and 1,3-DMBA, and two new experimental stimulants, octodrine and 1,4-DMAA. Of the four stimulants, only the stimulant octodrine has previously been marketed as an oral drug, and the recommended serving size found in our study was greater than twice the highest pharmaceutical dose previously available. In animal models, all four of these compounds have pressor effects. The FDA has warned that one of these stimulants, 1,3-DMAA, may pose cardiovascular risks, and it is possible that the other three stimulants may pose similar risks to consumers. Consumers should be warned about the presence of these four experimental stimulants in supplements. Until these stimulants are eliminated as supplement ingredients, we recommend that consumers avoid supplements labeled as containing 2-aminoisoheptane, 2-amino6-methylheptane, DMHA or Aconitum kusnezoffii. Physicians should remain alert to the possibility that patients may be inadvertently exposed to experimental stimulants when consuming weight loss and sports supplements.

October 12, 2018

Journal of the American Medical Association Network Open

"Unapproved Pharmaceutical Ingredients Included in Dietary Supplements Associated with US Food and Drug Administration Warnings"[18]

Findings:

In this quality improvement study, analysis of the US Food and Drug Administration warnings from 2007 through 2016 showed that unapproved pharmaceutical ingredients were identified in 776 dietary supplements, and these products were commonly marketed for sexual enhancement, weight loss, or muscle building. The most common adulterants were sildenafil for sexual enhancement supplements, sibutramine for weight loss supplements, and synthetic steroids or steroid-like ingredients for muscle building supplements, with 157 products (20.2%) containing more than 1 unapproved ingredient.

Meaning:

Potentially harmful active pharmaceuticals continue to be identified in over-the-counter dietary supplements.

Discussion:

The presence of pharmaceutically active ingredients in dietary supplements makes them unapproved drugs and represents an important public health concern. Of products that were found to be adulterated more than once, 19 (67.9%) had new drug ingredients reported in their second or third warning. This indicates that these products continue to be sold and are potentially dangerous even after FDA warnings. This is alarming, especially considering that the FDA is only able to test a portion of products available on the market.

Conclusions:

Dietary supplements are not subject to premarket approval for safety and effectiveness by the FDA and some have been found to contain undeclared drug ingredients. Of products found to be adulterated more than once, the majority were reported to contain new drug ingredients in subsequent warnings, indicating that adulterated dietary supplements continue to be an issue even after FDA action.

The active pharmaceutical ingredients identified in dietary supplements are present at unknown concentrations and have not been characterized as safe and effective by the FDA, making them unapproved drugs. These products have the potential to cause severe adverse health effects owing to accidental misuse, overuse, or interaction with other medications, underlying health conditions, or other drugs within the same dietary supplement. As the dietary supplement industry continues to grow in the United States, it is essential to further address this significant public health issue.

October 22, 2018

Journal of the American Medical Association Internal Medicine

"Prohibited Stimulants in Dietary Supplements After Enforcement Action by the U.S. Food and Drug Administration"[19]

The US Food and Drug Administration (FDA) is responsible for eliminating adulterated and potentially hazardous dietary supplements from the marketplace. The FDA uses a variety of enforcement actions, including public notices, to remove potentially hazardous ingredients. However, it is not known whether public notices are effective. We explored the effectiveness of the FDA's public notices issued between 2013 and 2016 targeting prohibited sympathomimetic stimulants in supplements. We analyzed supplements purchased in 2014 and the same brands purchased again in 2017 to determine the presence of prohibited stimulants before and after the FDA issued public notices.

In the Discussion Section:

To eliminate potentially hazardous supplements from the marketplace, the FDA recalls individual products and issues public notices regarding individual ingredients. The effectiveness of FDA recalls of individual products has been previously studied. One analysis found that 67% of brands subject to FDA recalls still on sale contained adulterants. Our current study explores the use of public notices targeting individual ingredients in supplements rather than individual products. Two findings are notable. First, the number of products that contained 1,3-DMAA, BMPEA, and oxilofrine decreased, but most supplements tested contained 1 or more prohibited stimulant, some up to 4 years after FDA action. Second, 1 stimulant was introduced only after FDA enforcement action. Future studies will be necessary to determine whether the FDA's public notices may, on occasion, inadvertently lead to the introduction of prohibited stimulants in supplements.

Our study has several limitations. It was small; we analyzed 12 brands of supplements at 2 time points during a 3-year period. Larger studies will be necessary to confirm our findings. Furthermore, we analyzed only 1 sample of each supplement, and stimulants might vary from batch to batch.

Despite these limitations, our study provides further evidence that a regulatory system that relies on post market enforcement activities is insufficient to ensure the safety of dietary supplements. Practitioners should advise patients that dietary supplements may contain prohibited stimulants.

December 2018

Journal of the American Medical Association

Editor's Note: "Regulating the Dietary Supplement Industry—The Taming of the Slew"[20]

The iconic image of the snake oil salesman, hawking his panaceas and elixirs, reminds us that the sale of unregulated medicinal products has been debated for more than a century. Interestingly, the origin of the term dates back to a decision rendered by the predecessor of the US Food and Drug Administration (FDA)—the Bureau of Chemistry—on Clark "the

Rattlesnake King" Stanley in 1916. Through chemical analysis, the bureau found that Stanley's snake oil, in fact, contained no snake oil at all but rather capsaicin, camphor, and turpentine. Hoping to make an example of him, federal prosecutors took Stanley to court for misbranding his product under the newly enacted Pure Food and Drug Act, ultimately fining him the lofty sum of $20. It is unclear what influence this had at the time, but 100 years later snake oil remains available as just one of a vast number of nutritional supplements marketed and sold without routine oversight.

October 12, 2018

Journal of the American Medical Association Network Open/ Public Health

"The FDA and Adulterated Supplements—Dereliction of Duty"[21]

A Commentary by Pieter Cohen, M.D.

The US Food and Drug Administration (FDA) plays an essential role in ensuring the safety of vitamins, minerals, botanicals, probiotics, amino acids, and glandular extracts sold as dietary supplements in the United States. While the FDA does not assess the safety of supplements prior to market, the agency is tasked with identifying and removing adulterated and hazardous supplements from the marketplace.

Adulteration of dietary supplements typically involves 1 of 2 patterns: economic adulteration, in which a less expensive ingredient is used in place of a more expensive ingredient listed on the label, or pharmaceutical adulteration, in which an active drug is included in a purportedly botanical supplement, for example, sildenafil in a "natural" sexual enhancement supplement. The FDA maintains a public database listing the brands of supplements it has identified as adulterated with drugs and the actions, if any, it has taken to remove the product from commerce.

An analysis of the FDA database of pharmaceutically adulterated supplements is the focus of a new study by Tucker and colleagues. The authors found that between 2007 and 2016 the FDA identified 746 brands of supplements adulterated with pharmaceutical agents. The adulterants included prescription medications such as sildenafil and fluoxetine, withdrawn medications including sibutramine and phenolphthalein, and unapproved drugs including dapoxetine and designer steroids. Twenty percent of the adulterated supplements contained 2 or more undeclared drugs, for example, weight loss supplements containing both an anorectic and a laxative. Most supplements adulterated with drugs were marketed as weight loss, sexual enhancement, or sports supplements—the same categories that epidemiologists have found to be responsible for a disproportionate number of the estimated 23 000 emergency department visits attributed to dietary supplements each year in the United States.

Given the potential public health risks of inadvertently ingesting unknown quantities of pharmaceutical drugs, once an adulterated supplement has been identified by the FDA, the agency frequently requests that the responsible firm voluntarily recall the product and, if the firm agrees, the agency publicizes the recall through email alerts and postings on its website. However, the effectiveness of voluntary recalls for supplements has been questioned. In one study, investigators found that many supplements previously subject to recalls remained on sale and were still adulterated with pharmaceutical drugs, sometimes years after the initial recall. In another study, consumers of a supplement subject to a voluntary recall were not aware of the recall and continued to purchase the product following the recall.

Despite their limited effectiveness, voluntary recalls are the most common approach used by the FDA to remove adulterated supplements from commerce. In the current study, the agency discovered 746 distinct supplements to be adulterated but announced voluntary recalls for only 360. Only 360 of 746 (48%) were recalled.

The database does not provide information as to why the FDA fulfilled its responsibilities less than half of the time, but it is possible that some firms might have refused to voluntarily recall their products. Warning letters may be used to nudge firms to recall supplements. In the current study, however, more than 140 firms were involved, but the FDA issued only 7 warning letters. The agency has other enforcement tools at its disposal when a firm does not agree to a voluntary recall, including mandating a recall (authority available since 2011 under the FDA Food Safety Modernization Act) or making a referral to the Department of Justice. Tucker and colleagues1 found that the agency seldom uses these enforcement tools: the FDA reported no mandatory recalls and only 1 Department of Justice investigation in response to the 746 brands of adulterated supplements.

This new evidence is consistent with prior research that has highlighted major deficiencies in the FDA's regulation of supplements. In a similar study published in 2013, Harel and colleagues found that the FDA identified 332 brands of supplements adulterated with pharmaceutical agents during the 9-year period from 2004 to 2012 but only 222 brands (67%) were recalled. In another investigation from 2013, the FDA's analytical chemists uncovered a mixture of synthetic compounds, including an amphetamine analog, β-methylphenylethylamine (BMPEA), in weight loss and sports supplements. The FDA did not inform consumers or issue warning letters. An independent study describing the FDA's inaction was published 2 years later, and only then did the FDA begin to take steps to remove the supplements containing BMPEA from the market.

This pattern is currently repeating itself—the FDA has not warned consumers about additional stimulants discovered in weight loss and sports supplements. My colleagues and I informed the FDA in early 2017 that we had identified 2 experimental stimulants, 1,4-dimethylamylamine and octodrine, in dietary supplements. One stimulant has never been approved by the FDA for use in humans, and the other was approved for use by inhalation in the 1940s but has since been removed from the US market. Neither stimulant has ever been FDA approved for oral consumption. Our research has since been confirmed by FDA-funded investigators, yet as of September 2018 the FDA has not taken any regulatory action to remove these synthetic stimulants from commerce or warn consumers about the novel adulterants.

To counter the perception of regulatory inertia, FDA officials have emphasized their work to eliminate the stimulant 1,3-dimethylamylamine (1,3-DMAA) from supplements. The sympathomimetic 1,3-DMAA was originally introduced by Eli Lilly & Co in the 1940s as a nasal decongestant to compete with amphetamine marketed by Smith, Kline and French. By the 1970s, 1,3-DMAA had been withdrawn from the US markets, but it reappeared in the 2000s as a replacement for ephedra in sports and weight loss supplements; by 2012 the stimulant was available in more than 200 brands of supplements. The World Anti-Doping Agency banned the stimulant in sport in 2009. In 2011, Health Canada banned 1,3-DMAA from supplements and the US Department of Defense removed 1,3-DMAA supplements from military bases due to safety concerns. The stimulant received prominent media attention as potentially contributing to strokes and deaths of US troops. Only in 2012 did the FDA finally begin to use its full enforcement powers, including warning letters, product seizures, and mandatory recalls, to remove the stimulant from supplements.

More than FDA action will be required to ensure that all adulterated supplements are effectively and swiftly removed from the market. Congress would need to reform the Dietary Supplement Health and Education Act of 1994. One practical change would be to require firms to register supplements with the FDA prior to sale and Congress could provide the FDA with more effective enforcement tools such as immediately revoking a product's registration if a supplement is found to be adulterated with pharmaceutical drugs. In the meantime, the process that the FDA is required to follow to remove supplements from the marketplace will remain cumbersome and time-consuming; nevertheless, the agency's failure to aggressively use all available tools to remove pharmaceutically adulterated supplements from commerce leaves consumers' health at risk.

April 1, 2019

Alimentary Pharmacology & Therapeutics

"Severe and Protracted Cholestasis in 44 Young Men Taking Bodybuilding Supplements"[22]

Abstract:

Forty-four males (mean age 33 years) developed liver injury with median latency of 73 days. Seventy-one percent were hospitalized.

From Section 3.5 of the Paper—Chemical Analysis of Bodybuilding Supplement Components:

- Thirty-three bodybuilding products were collected.
- At least, one anabolic steroid was identified in 14 bodybuilding supplements.
- Thirteen of the steroids identified were DEA controlled substances.

Final Comments

The information provided in this book illustrates the needlessness of supplement use for most people who maintain even reasonable dietary habits. Based upon this information, it is the author's belief that any athlete or consumer who has become aware of the objective information about the possible hazards of many supplements purported to enhance one's weight loss efforts, physique, or physical performance, and continues to consume them, are literally playing Russian roulette with their health.

Notes

1. www.statnews.com/2017/01/10/supplement-harvard-pieter-cohen/
2. Hans Geyer, Maria Parr, Karsten Koehler, Ute Mareck, Wilhelm Schänzer, and Mario Thevis, Nutritional supplements cross-contaminated and faked with doping substances. *Journal of Mass Spectrometry* (2008), Vol. 43, pp. 892–902.
3. Z. Harel, S. Harel, R. Wald, M. Mamdani, C.M. Bell, The frequency and characteristics of dietary supplement recalls in the United States. *JAMA Internal Medicine* (2013), Vol. 17, No. 10, pp. 929–30.
4. Y. El Sherrif, et al., Hepatotoxicity from anabolic androgenic steroids marketed as dietary supplements: contribution from ATP8B1/ABCB11 mutations? *Journal of Liver International* (September 2013), Vol. 33, No. 8, pp. 1266–70.
5. Pieter A. Cohen, John C. Travis, Bastiaan J. Venhuis. A methamphetamine analog (N, α-diethyl-phenylethylamine) identified in a mainstream dietary supplement. *Drug Testing and Analysis* (2013), pp. 805–7.

6. Katz, Mitchell, How can we know if supplements are safe if we do not know what is in them? *JAMA Internal Medicine* (2013),Vol. 173, p. 928.

7. harvardpublichealthreview.org/how-americas-flawed-supplement-law-creates-the-mirage-of-weight-loss-cures/

8. Victor Navarro, Huiman Barnhart, Herbert Bonkovsky, Timothy Davern, Robert Fontana, Lafaine Grant, Rajender Reddy, Leonard Seeff, Jose Serrano, Averell Sherker, Andrew Stolz, Jayant Talwalkar, Maricruz Vega, and Raj Vuppalanchi, Liver injury from herbals and dietary supplements in the US drug-induced liver injury network. *Hepatology* (August 25, 2014),Vol. 60, No. 4, pp. 1399–408.

9. www.fda.gov/drugs/medication-health-fraud/public-notification-sport-burner-contains-hidden-drug-ingredient

10. www.nejm.org/doi/full/10.1056/NEJMp1315559

11. www.nejm.org/doi/full/10.1056/NEJMsa1504267

12. https://onlinelibrary.wiley.com/doi/pdf/10.1002/dta.1899

13. www.justice.gov/opa/pr/justice-department-and-federal-partners-announce-enforcement-actions-dietary-supplement-cases

14. www.fda.gov/consumers/consumer-updates/caution-bodybuilding-products-can-be-risky

15. https://www.fda.gov/news-events/fda-brief/fda-brief-fda-warns-against-using-sarms-body-building-products

16. www.fda.gov/drugs/drug-safety-and-availability/fda-issues-warning-about-body-building-products-labeled-contain-steroid-and-steroid-substances

17. Pieter Cohen, John Travis, Peter Keizers, Patricia Deuster, and Bastiaan Venhuis, Four experimental stimulants found in sports and weight loss supplements: 2-amino-6-methylheptane (octodrine), 1,4-dimethylamylamine (1,4-DMAA), 1,3-dimethylamylamine (1,3-DMAA) and 1,3-dimethylbutylamine (1,3-DMBA). *Clinical Toxicology* (November 2017),Vol. 56, No. 6, pp. 1–6.

18. J. Tucker, T. Fischer, L. Upjohn, D. Mazzera, M. Kumar, Unapproved pharmaceutical ingredients included in dietary supplements associated with US Food and Drug Administration warnings. *JAMA Network Open* (October 12, 2018),Vol. 1 No. 6.

19. P.A. Cohen, A. Wen, R. Gerona, Prohibited stimulants in dietary supplements after enforcement action by the US Food and Drug Administration. *JAMA Internal Medicine* (2018),Vol. 178, No. 12, pp. 1721–3.

20. M. Incze and M.H. Katz, Regulating the dietary supplement industry: the taming of the slew. *JAMA Internal Medicine* (2018),Vol. 178, No. 12, p. 1723.

21. https://jamanetwork.com/journals/jamanetworkopen/fullarticle/2706489

22. https://onlinelibrary.wiley.com/doi/10.1111/apt.15211

10
ANTI-INFLAMMATORY, DETOX, AND KETOGENIC DIETS

Anti-Inflammatory Diets

Just about every health guru on the planet, as well as many respected science organizations, has their advice regarding purported inflammatory and anti-inflammatory foods and their purported role in the prevention or development of a myriad of diseases. But is this advice based upon a clear understanding, or a premature assumption of the biological reactions which are occurring with various foods after they have been digested?

Since this issue is far too expansive, thus making it virtually impossible to isolate all of the incredulous comments regarding anti-inflammatory diet plans, I am going to use a group which many would normally consider as being on "top of the food chain," so to speak, regarding sound nutrition science advice. Most individuals would agree that the following group should retain a keen eye for data interpretation and the ability to draw warranted conclusions vs. the tidal wave of unjustified rubbish in the uneducated culture at large. So, I will use the Department of Nutrition, Harvard T.H. Chan School of Public Health, as my example here.

On November 7, 2018, *Harvard Health Publishing—Trusted Advice for a Healthier Life*—which is rather an ironic comment, as you will see—published "Foods That Fight Inflammation."[1] In this article, they state the following:

> Doctors are learning that one of the best ways to reduce inflammation lies not in the medicine cabinet, but in the refrigerator (they are not referring to the ice). By following an anti-inflammatory diet, you can fight off inflammation for good.

Later in the same newsletter, it is stated:

> Choose the right anti-inflammatory foods, and you may be able to reduce your risk of illness. Consistently pick the wrong ones, and you could accelerate the inflammatory disease process.

The article goes on to identify the common culprits of "inflammation," as:

- **refined carbohydrates**, such as white bread and pastries
- **French fries** and other fried foods
- **soda** and other sugar-sweetened beverages
- **red meat** (burgers, steaks) and processed meat (hot dogs, sausage)
- **margarine**, shortening, and lard

And anti-inflammatory foods as:

- **tomatoes**
- **olive oil**
- **green, leafy vegetables**, such as spinach, kale, and collard greens
- **nuts**, like almonds and walnuts
- **fatty fish**, like salmon, mackerel, tuna, and sardines
- **fruits** such as strawberries, blueberries, cherries, and oranges

Now let's look at the evidence. After the ingestion of various types of purported inflammatory foods, such as refined carbohydrates, French fries, sodas, etc., as mentioned, certain molecules (or biomarkers) known to be associated with the inflammatory process increase in concentration in the blood after the foods are consumed. Now, the false assumption is that this temporary increase of biomarkers normally associated with inflammation is a negative issue and not a normal, temporary physiological response after the ingestion of certain foods—especially high glycemic index foods. Now, are these known inflammatory biomarkers which are being released an indication of inflammation or an indication of another, yet unidentified, role these biomarkers may have unrelated to inflammation?

Marc Donath, M.D., is Professor and Head of the Clinic for Endocrinology, Diabetes and Metabolism at the University Hospital of Basel, Switzerland. His research focuses on the mechanisms and therapy of decreased insulin production in type 2 diabetes. He has shown that IL-6, one of the biomarkers of inflammation widely used to condemn certain foods as "inflammatory," as the ones mentioned in the preceding list, regulates another molecule production called GLP-1. So, is this "inflammatory" biomarker, IL-6, being mischaracterized as a negative reaction to the food being ingested, or just a separate normal biological role for this molecule?

GLP-1, which is stimulated by the "inflammatory" biomarker, is responsible for stimulating the B-cells of the pancreas to secrete more insulin, all of which is dependent upon blood sugar levels after a meal. GLP-1 also suppresses the hormone glucagon, which, if not suppressed, would stimulate the release of even more sugar into the blood. Thus, this reaction is a good thing and not a bad thing. The inflammatory biomarker initiates the initial necessary steps to maintain normal blood sugar levels after the ingestion of various high glycemic index foods. GLP-1 also induces satiety in the hypothalamus, which, if you pay attention to this signal from your brain, it will help prevent overindulgence and obesity. Again, so far this is all normal physiology, initially stimulated by the inflammatory biomarker. The inflammatory biomarker is simply an indication of your body's normal physiological response to blood sugar levels and based upon Dr. Donath's research, is a necessary step of normal physiological reactions to maintain normal blood sugar homeostasis. So, what does all this mean in simplified functional terms?

If you consume any food which may stimulate significant increases in blood sugar levels, which would be labeled as an "inflammatory food," such as potato products, refined carbohydrates, sodas, etc., what is going to happen? Your blood sugar will rise significantly. Then, the normal physiological response of the body is to secrete the insulin to control the level of blood sugar and keep it within normal ranges. Insulin will do this by stimulating the liver and muscle tissue to take up the excessive sugar. The two sites can either store it if possible as glycogen, or, in the liver's case, biosynthesize fatty acids from the excess sugar and store it in fat tissue. This is all perfectly normal metabolism of excessive calories, and the biomarkers for inflammation which have received so much negative attention are simply assisting in this process by stimulating the necessary initial steps of it. Increased blood sugar stimulates the inflammatory biomarker, which then stimulates the GLP-1 molecule, which then stimulates the pancreatic beta cells to secrete insulin. This is not chronic inflammation, but a temporary secretion of a biomarker for inflammation, which obviously has other biological roles, to elicit the normal metabolic functions to control blood sugar levels.

Here is an email I sent to Dr. Donath on February 2, 2019 regarding this issue:

It appears, the short-term inflammatory marker response is essentially a normal physiological response to increased blood sugar levels or excess calories, and the need to direct the appropriate hormonal responses to metabolize the sugars or excess calories. Is this correct or incorrect? Would you consider the current consumer and media reaction to the inflammatory marker response to food, similar to the misguided response to normal free radical production? In other words, under many conditions, such as post-exercise and during various immune responses, the increase free radical production is a normal metabolic response which acts as a signaling mechanism for various biological adaptations which need to take place, such as signaling the muscle tissue to adapt to the exercise load. Suppressing this free radical response hinders the adaptation. So, even though free radicals are often associated with negative biological activity by most consumers, during many physiological situations, their response is simply a signaling mechanisms for further normal metabolic reactions in many situations. Is this also true for the markers of inflammation postprandial?

His response, on February 10, 2019:

Overall, I agree with your thinking.

To support all of this, in March 2017, Dr. Donath and his colleagues published their research findings in *Nature Immunology*, "Postprandial Macrophage-Derived IL-1B Stimulates Insulin, and Both Synergistically Promote Glucose Disposal and Inflammation."[2] The discussion section of this paper states that "both insulin and IL-1B (biomarker for inflammation) regulated whole body glucose disposal by promoting glucose uptake in muscle and fat and fueled the immune system by stimulating the uptake of glucose into the immune-cell compartment." Also, their final comments were: "collectively, our findings have shown that IL-1B, a master regulator of inflammation, and insulin, a key hormone in glucose metabolism, promoted each other. Both had potent effects on glucose homeostasis and on the activity of the immune system."

On January 16, 2017, *Science Daily* covered the same research of Dr. Donath in its review "Every Meal Triggers Inflammation."[3] In this review, the following points are made:

- When we eat, we do not just take in nutrients—we also consume a significant quantity of bacteria. The body is faced with the challenge of simultaneously distributing the ingested glucose and fighting these bacteria. This triggers an inflammatory response that activates the immune systems of healthy individuals and has a protective effect.
- This inflammation does have some positive aspects. In healthy individuals, short-term inflammatory responses play an important role in sugar uptake and the activation of the immune system.

The Bottom Line

Certain foods, due to their effect on blood sugar levels, precipitate the release of various molecules which are also associated with inflammation. However, this does not mean these foods are triggering inflammation or disease. Their presence is just a normal response and dual role these molecules have to assist in the maintenance of normal blood sugar homeostasis. Consuming refined carbohydrates, French fries, or sodas is not going to cause inflammation. Their excessive consumption will certainly be reflective of an LSD diet (Lousy Stinking Diet), which is one low in plant-based foods and the subsequent absence of thousands of healthy plant chemicals, and a likely similar overall poor lifestyle habit. So of course, these individuals are going to have significant negative health issues, but it certainly is not going to be related to the occasional or limited daily use of the so-called "inflammatory" foods. To suggest so is junk science and a lack of common sense.

Detox Diets

There are a variety of various recommended body detox regimens to cleanse what you purportedly "polluted" your body with. It could be in the form of a pair of socks, which will allegedly extract toxins through your feet, diets that will assist the efforts of your liver, or a good old-fashioned colon cleanse to scrub your bowels.

There is no question that overindulgence, a.k.a. gluttony, leaves us feeling as if we have "polluted" ourselves, but is there any real science behind detoxing, or cleansing—or is it a placebo effect? Is it science or quackery? Let's look.

We are exposed to "toxins" daily, and they come in the form of both exotoxins, the ones we are exposed to through the air, drinking and eating, as well as endotoxins, which are the normal byproducts of our daily metabolism. However, just how much of an impact do these molecules have on us that our bodies are not already well-equipped to handle under normal circumstances and exposure levels?

Here are a few of the mechanisms put in place to assist your efforts to maintain your health.

The cell: Normal cell metabolism produces a wide variety of waste products, some of which can be highly toxic if there were no means of disposing of them. Within each cell, there are specialized vesicles called lysosomes and peroxisomes, which are essentially the cells recycling and waste disposal center for cellular debris. Each lysosome has an estimated 60 different enzymes that digest cell waste, viruses, and bacteria, then dispose of them via the liver and kidneys for removal. Additionally, every cell produces an important detoxifying compound called glutathione. This compound has significant detoxifying roles and its concentration within the cell is increased with exercise, which is one reason regular exercisers have stronger immune systems.

Liver: One of the many functions of the liver is to detoxify external chemicals (exotoxins) after they have been absorbed from food, water, or air, as well as internally produced chemicals (endotoxins), which are normal every-day metabolic byproducts of cell metabolism, such as nitrogen from excess protein intake. Any excess nitrogen in the body is converted to ammonia, which if left unchecked would cause brain damage. However, the liver readily transforms the ammonia into urea, which is then eliminated through the kidneys with your urine. The liver also makes bile, which contains some metabolic waste, such as bilirubin. The liver does not store toxins, but reduces unwanted chemicals into water-soluble compounds, which can then be eliminated through the kidneys.

Intestines: Most people think of their intestines as a porous tube which will essentially allow the absorption of anything that passes through it. This is false; it can be very selective. As an example, iron is an important mineral for both males and females. However, free iron in the blood is highly toxic to any tissue or organ it encounters due to its extremely high oxidative capacity. So, iron must be bound to a limited number of iron-binding proteins in your blood. Therefore, your intestines will be very selective as to how much iron is absorbed. An excess would be life-threatening. Intestinal flora, the good bacteria, inhibits the growth of unfriendly bacteria. Additionally, those who would like you to believe you may be experiencing a build-up of some mysterious intestinal wall sludge, which may be preventing the normal function of your intestines, have no understanding of the extensive enzyme and flora activity in your intestines. All one must do is view a normal colonoscopy from any friendly local gastroenterologist, which will visually refute the sludge build-up claims.

Kidneys: The kidneys can filter over 30 gallons of blood per day and readily separate the waste, which is then eliminated in the urine, from the compounds which should be recycled, such as nutrients. As an example, we often hear scary stories of "studies" that have shown that those individuals who consumed produce which may have had a pesticide residue on it have higher rates of pesticides in their urine. Sounds toxic, but this reflects the normal functioning of the liver and kidney working in concert with each other to eliminate compounds you have no need for.

Lungs: The lungs expel carbon dioxide, which is a waste product of many metabolic functions such as muscle activity, metabolism of nutrients, etc. As the level of carbon dioxide rises in your blood, your respiration rate will also increase to assist in eliminating it. This is the main driving force of your rapid breathing during exercise.

The Bottom Line

This is by no means is a complete list of your body's ability to maintain healthy chemical homeostasis and neutralize harmful chemicals, but it should serve to give you a good idea as to how the functions of your physiology illustrate why "detox" or "body cleansing" diets are deceptions. You may feel better after your "detox," but this is nothing more than a placebo effect. Your body naturally detoxifies itself, and yes, it is not perfect, and people do get sick and die, but this has nothing to do with the need of a "cleanse." If you really want to detox physically, simply make wiser food choices and exercise.

Ketogenic Diets

There are a considerable number of advocates of the low-carb diet regimen: some related to treating various medical conditions such as type 2 diabetes and high blood pressure, as well as for weight loss and obesity. The physiological reasons for this are logical, as will be pointed out, but the physiological reasons to adhere to this dietary protocol do not hold up, and neither is it logical, based upon a number of factors.

The overriding point here which needs to be understood is what basic physiological needs are, or are not being met by a ketogenic diet, and, if those physiological needs are not being met, what are the potential long-term negative effects on your health and physical performance?

The following facts are known to be true:

- Glucose is the main, or preferred, energy source for several tissues, such as the brain, muscle tissue, phagocytic cells of the immune system, central nervous system, red blood cells, digestive microflora, etc. I pointed out in Chapter 6 the work of Dr. Nieman regarding how a low-carbohydrate diet negatively affects the immune system. To refresh your memory of this issue, return to the "Decreased Immunity" section under "Variable 1—A Carbohydrate-Deficient Diet" section of that chapter.
- Carbohydrates are the only source of fiber to maintain a healthy digestive tract/colon in order to prevent diverticulosis and colon cancer (see the appropriate images later in this chapter). Without the fibers from such food items as fruit, grains, legumes, and vegetables, cancer-protective bacteria in our colon fail to proliferate. Various microbes of good bacteria in the intestinal tract utilize fibers and digested resistant starches to feed the good gut bacteria, which in turn help us digest food, enhance our immune system, and destroy pathogens. In 2011, the *American Journal of Clinical Nutrition* published "High-Protein, Reduced-Carbohydrate Weight-Loss Diets Promote Metabolite Profiles Likely to Be Detrimental to Colonic Health."[4] In this study, the researchers assessed the effects of various dietary protocols on the gut microbiota and their derived metabolites that may influence the health of the colon. In this study, 17 obese men, those most likely to be attracted to the ketogenic diet, were provided with either four weeks of

a high-protein diet or four weeks of a moderate carbohydrate diet. Their conclusion:

> After 4 wk., weight-loss diets that were high in protein but reduced in total carbohydrates and fiber resulted in a significant decrease in fecal cancer-protective metabolites and increased concentrations of hazardous metabolites. Long-term adherence to such diets may increase risk of colonic disease.[5]

- Carbohydrates are necessary for normal fat metabolism, which is, of course, the basis of the ketogenic diet. The ketogenic diet forces the body to abnormally produce ketone bodies (acetone, acetoacetate, and B-hydroxybutyrate) in much larger amounts than normal when carbohydrates are unavailable, and use these bodies as an energy source. This, of course, is appealing to many advocates of weight loss and control diets, because it will produce a reliance on fatty acids from fat mass as the sole energy source. This will have a short-term advantage for weight loss purposes, which has been demonstrated, but what are the long-term consequences?
- On September 1, 2018, researchers from the Cardiovascular Division, Department of Medicine at Brigham and Women's Hospital in Boston, published "Dietary Carbohydrate Intake and Mortality: A Prospective Cohort Study and Meta-Analysis."[6] They found that mortality rates increase at either end of the spectrum of carbohydrate intakes and that the preferred level of carbohydrates in this non-athletic population group was 50–55%. Low-carbohydrate was defined as below 40% of total energy intake, substantially higher than the ketogenic advocates, and high was defined as greater than 70%.
- Both acetoacetate and beta-hydroxybutyrate are acidic, so if these ketone bodies are too high in the blood, overwhelming the alkaline reserve of the blood buffering capacity, the blood's pH will drop resulting in ketoacidosis. Even mild acidosis may impair anaerobic or high-intensity exercises, because the acidosis impairs enzymatic reactions in muscle tissue. This was demonstrated in a study published in the *Journal of Sports Medicine and Physical Fitness* April 4, 2018, "Low-Carbohydrate, Ketogenic Diet Impairs Anaerobic Performance in Exercise-Trained Women and Men: A Randomized-Sequence Crossover Trial."[7]
- Ketoacidosis can produce some of the following symptoms if the ketones in the blood are excessive: excessive thirst, frequent urination, abdominal pain, nausea and vomiting, weakness and fatigue due to the lack of glucose, shortness of breath, etc.
- The elimination of any plant-based food, such as fruit, starchy vegetables, grains, etc., which may have a higher than "desired" carbohydrate content for the ketogenic diet enthusiast, will eliminate literally thousands of

phytochemicals from the diet. This will have major, negative long-term health consequences, such as increased rates of colon cancer. As an example, phytochemicals flavonoids, carotenoids, alkylresorcinols, phenolic acids, resveratrol, saponins, and tannins, are just a few examples of compounds which are loaded in fruits and grains and which have anticarcinogenic and immunological functions. These beneficial food items would be low or deficient in many ketogenic diets.

- There are a few studies which do support some of the short-term claims of ketogenic advocates, but they are of short duration with small sample sizes, and are mixed and contradictory in their reported results. The small sample sizes and short duration studies make it impossible to really assess the long-term purported benefits apart from fat mass loss. The *Journal of International Society of Sports Nutrition*, on July 12, 2017, published "Ketogenic Diet Benefits Body Composition and Well-Being But Not Performance in a Pilot Case Study of New Zealand Endurance Athletes."[8] After ten weeks on the ketogenic diet, all athletes naturally increased their ability to utilize more fatty acids as a fuel source, which would be expected, but they experienced "a decreased ability to undertake high intense bouts of exercise," which applies to most sporting events. This would be expected due to the lack of glycogen in muscle tissue while on a ketogenic diet. No immediate fuel source, no power.

In April 2018, the journal *Metabolism* ran an article headlined "Keto-Adaptation Enhances Performance and Body Composition Responses to Training in Endurance Athletes,"[9] but stated in the results section that "there was no significant change in performance of the 100km time trial between groups." So, this is contradictory.

It is true that there are going to be some short-term improved health markers on a ketogenic diet, but these are only brief indicators of what one would expect from such a diet regimen. It is common sense that if you eliminate most carbohydrates from your diet, your blood sugar will be lower. For diabetics who are finding it very difficult to control their blood sugar levels, this may be a short-term goal for better control. For weight loss, it is also commonsense that if you force your body to rely totally on fat as a fuel source, you are going to lose more fat mass over a period of four weeks or longer than someone expending the same amount of daily energy would achieve. However, it is clear that that due to the loss of muscle glycogen, intense activities requiring 80% to 100% of maximal efforts are going to be severally limited. It is this activity which enhances muscle growth, the very tissue which will enable better control of blood sugar levels in the long term, as well as weight loss and control. Some will counter that the data illustrates that those following a ketogenic diet have more lean body mass after following the diet. This is also true, but the definition of lean body mass is anything other than fat mass. So of course, with greater fat loss as well as significant fluid loss, anyone on a ketogenic diet will have an improved lean body mass—but this does not necessarily equate to muscle mass. To settle this point, there would have to be long-term studies illustrating that those athletes who follow a ketogenic diet have been able to achieve greater muscle gains and strength performance over those athletes who followed the traditional Mediterranean-type diet. In Chapter 6, I provided the 2019 study published in the *Journal of the American College of Nutrition*, which illustrated the 6% improvement in a 5-km run time after just four days of switching from a typical Western diet to the Mediterranean-type diet.

Finally, it is also common sense that blood pressures would improve on a ketogenic diet due to the weight loss and the diuretic effect of the protein being used as an energy source. I illustrate this mechanism in Misconception 2 in Chapter 6. As protein is catabolized, the resulting increased urea formation and delivery to the kidneys must, obviously, be excreted. However, as I state in Chapter 6, the kidneys must dilute the higher concentration of urea now found in the urine, resulting in an increased fluid loss, as pointed out in Chapter 6, resulting in as much as a four to fivefold increase in urine volume in some individuals consuming in excess of 2.0g/kg/day. Essentially, the excessive protein could produce a diuretic effect. This diuretic effect will illicit additional "weight" loss, but not necessarily fat mass—just water weight. And of course, this will assist in better blood pressure control for some.

The Bottom Line

Yes, it is true that a ketogenic diet will likely result in some positive outcomes for blood sugar control and blood pressure, as well as weight loss, due to the obvious short-term physiological adaptations which take place which will benefit this population group. However, these results are short-sighted, in my opinion, and vanity has become the primary motivating factor for most who follow this regimen rather than long-term health. Ironically, due to the long-term negative impact a diet low in carbohydrates (phytochemicals) will likely have, ketogenic diet enthusiasts appear willing to follow a dietary strategy loaded with detour and cautionary signs warning them of the potential long-term consequences for a brief period of kudos on their physical appearance.

To sum up the foolishness of this diet, read the information I provided in Chapter 8 under Misconception 4, where I discuss the work done by Cornell University researchers illustrating how the synergistic effect of the phytochemicals in apples had such a dramatic effect on the prevention of proliferation of colon cancer cells. After that review, ask yourself the following question. Below are three images, one of diverticulosis and diverticulitis (Image 10.1), and the other of colon cancer (Image 10.2). Does the ketogenic diet put you at greater risk for these disease conditions?

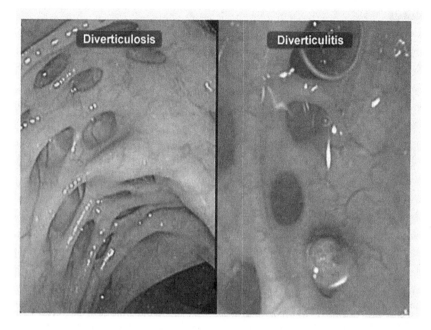

IMAGE 10.1 Diverticulosis and diverticulitis.

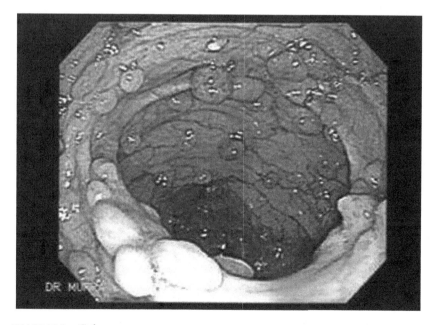

IMAGE 10.2 Colon cancer.

Notes

1. www.health.harvard.edu/staying-healthy/foods-that-fight-inflammation
2. M. Donath, et al., Postprandial macrophage-derived IL-1B stimulates insulin, and both synergistically promote glucose disposal and inflammation. *Nature Immunology* (January 16, 2017), Vol. 18, pp. 283–92.
3. Every meal triggers inflammation: Short-term inflammatory responses play a key role in sugar uptake and activation of immune system. *Universität Basel, ScienceDaily* (January 16, 2007). www.sciencedaily.com/releases/2017/01/170116121912.htm
4. Wendy R Russell, Silvia W Gratz, Sylvia H Duncan, Grietje Holtrop, Jennifer Ince, Lorraine Scobbie, Garry Duncan, Alexandra M Johnstone, Gerald E Lobley, R John Wallace, Garry G Duthie, Harry J Flint, High-protein, reduced-carbohydrate weight-loss diets promote metabolite profiles likely to be detrimental to colonic health, *The American Journal of Clinical Nutrition* (2011), Vol. 93, No. 5, pp. 1062–72, https://doi.org/10.3945/ajcn.110.002188
5. Ibid, p. 1062.
6. www.thelancet.com/journals/lanpub/article/PIIS2468-2667(18)30135-X/fulltext
7. Kymberly A. Wroble, Morgan N. Trott, George G. Schweitzer, Rabia S. Rahman, Patrick V. Kelly, Edward P. Weiss, Low-carbohydrate, ketogenic diet impairs anaerobic exercise performance in exercise-trained women and men: a randomized-sequence crossover trial. *The Journal of Sports Medicine and Physical Fitness* (2018), Vol. 59, No. 4, pp. 600–7.
8. C. Zinn et al. Ketogenic diet benefits body composition and well-being but not performance in a pilot case study of New Zealand endurance athletes. *Journal of the International Society of Sports Nutrition* (July 12, 2017), Vol. 14, No. 22.
9. Fionn T. McSwiney et al., Keto-adaptation enhances exercise performance and body composition responses to training in endurance athletes. *Metabolism - Clinical and Experimental* (April 2018), Vol. 81, pp. 25–34.

11

GENETICALLY ENGINEERED FOOD

"Frankenfood" or Immensely Beneficial Technology?

On November 19, 2018, the Pew Research Center published its findings regarding Americans' opinion over the health effects of genetically modified (GM) foods.[1] They found that about half of US adults (49%) considered GM ingredients to be worse for one's health than those foods which contain no GM ingredients. It was also found this feeling was more likely to be held by women than men. Forty-four percent felt GM foods had no impact on one's health, and 5% felt they were better for you. The percentage of the population believing GM foods negatively affect one's health has increased by 10% since 2016. According to Pew Research, "People with low science knowledge tend to express more concern about the health risk from these food groups compared with those high in science knowledge."[2]

There is a clear analogy here which fits the purpose of this book. Throughout this book, I make it very clear that various groups have absolutely no financial incentive to educate consumers, but, instead rely on consumer misunderstandings and science illiteracy in order to exploit them. Whether it's the organic food industry, the supplement industry, the chemophobia issue, weight loss, or genetically modified organisms (GMOs), the pattern is always the same: motivate the consumer to agree with you by instilling an unwarranted fear using pseudoscience or junk science. This is a highly effective marketing strategy, but it relies on consumer ignorance. This is exactly what I have experienced in the classroom for 15 years. At the beginning of my college nutrition course, all students answer a brief series of questions regarding their thoughts on various popular health and lifestyle-related topics, as discussed throughout this book. The trend is always the same: the course begins with the majority of students embracing the false narratives for organics, anti-GMO, chemophobia, etc.; this list is long. However, as the course progresses and students are exposed to the flip side of the narrative that they have been indoctrinated with, the vast majority of them will flip their opinions and the prior fears or concerns they embraced at the beginning of the course are no longer an issue.

The need for a more objective narrative on GMOs was published in *Critical Reviews in Biotechnology* in 2014, titled "An Overview of the Last 10 Years of Genetically Engineered Crop Safety Research."[3] The review stated:

> The technology to produce genetically engineered (GE) plants is celebrating its 30th anniversary and one of the major achievements has been the development of GE crops. The safety of GE crops is crucial for their adoption and has been the object of intense research work often ignored in the public debate. We have reviewed the scientific literature on GE crop safety during the last 10 years, built a classified and manageable list of scientific papers, and analyzed the distribution and composition of the published literature. We selected original research papers, reviews, relevant opinions and reports addressing all the major issues that emerged in the debate on GE crops, trying to catch the scientific consensus that has matured since GE plants became widely cultivated worldwide. The scientific research conducted so far has not detected any significant hazards directly connected with the use of GE crops; however, the debate is still intense. An improvement in the efficacy of scientific communication could have a significant impact on the future of agricultural GE. Our collection of scientific records is available to researchers, communicators and teachers at all levels to help create an informed, balanced public perception on the important issue of GE use in agriculture.

As illustrated in the first two chapters of this book, most consumers obtain their "science" information from unreliable sources, which partly explains the wide disparity with what good scientific practices reveal regarding GMOs. This is very understandable. When you are routinely exposed to a negative narrative about anything, you tend to embrace that narrative as fact. In reality however, it is just an agenda-driven, one-sided storyline overloaded with emotionally charged personal views and pseudoscience. Unfortunately, the negative narrative which surrounds GMO food production has become a food marketing bonanza for many. All one must do to understand this obsession is to walk up and down any grocery store aisle and notice all the products that claim they are purportedly GMO free; this includes items which could not possibly have a GMO ingredient in it anyway. Now, is this fear of GMOs based on facts, or is the anti-GMO issue just another baseless fear? Should we move forward wisely with the technology to see if the technology will allow us to overcome some significant issues in producing the necessary food for billions of people, or should we vacate the technology and naïvely just hope for the best?

Historically, for thousands of years, farmers have bred (genetically modified) plants and animals to produce various desired traits they wanted, to enhance yield, quality, pest resistance, etc. Now, through genetic engineering, we have a technology that can simply speed up the once tedious and lengthy task of hybridization and achieve the same benefits which had taken literally decades to achieve before.

Consider some points regarding GE crops provided by Henry Miller and Gregory Conko in *Policy Review*'s "The Rush to Condemn Genetically Modified Crops," published by the Hoover Institution of Stanford University.[4] Mr. Miller is the Robert Wesson Fellow in Scientific Philosophy and Public Policy at Stanford University's Hoover Institution; he was the founding director of the FDA's Office of Biotechnology. Mr. Conko is a senior fellow at the Competitive Enterprise Institute.

- Genetic modification refers to sophisticated gene-splicing techniques in which genes are moved around precisely and predictably. This is unlike cruder cross breeding techniques which consumers have been accustomed to for decades, such as the consumption of a tangelo, which is a cross between a tangerine and a grapefruit.
- Every major scientific and public health organization such as the American Medical Association and the National Academy of Sciences, and many more, have concluded that gene splicing produces foods are as safe if not safer than conventional ones (such as cross breeding potatoes) to increase their virus resistance.
- There has not been one single confirmed adverse reaction or ecosystem disrupted with the use of GM crops which have been cultivated on more than 3 billion acres worldwide with the consumption of approximately 3 trillion servings in North America alone.
- The reduced fungal toxins alone make them safer for both consumers and livestock, especially in poorer countries.

In May 2016, the National Academy of Sciences, Engineering, and Medicine convened a committee to review the

claims and research that extol both the benefits of and risks posed by GE crops and food that have created a confusing landscape for the public and policymakers. Using evidence accumulated over the last two decades, this report assesses purported negative effects and purported benefits of currently commercialized GE crops. The report also assesses emerging genetic engineering technologies, how they might contribute to future crop improvement, and what technical and regulatory challenges they may present. To carry out its task, the report's authoring committee delved into the relevant literature (more than 900 research and other publications), heard from 80 diverse speakers at three public meetings and 15 webinars, and read more than 700 comments from members of the public to broaden its understanding of issues surrounding GE crops.[5]

The final report is 407 pages long, but a summarized version is available.[6]

The summarized "Report in Brief: Genetically Engineered Crops: Experiences and Prospects" points out the following environmental and crop production benefits:

- Crops which have been modified to increase the production of their own naturally occurring insecticide have reduced crop losses (increased yield per acre) and reduced the use of pesticides.
- Crops which have been modified to resist glyphosate, an unnecessarily maligned weed killer with the use of junk science, have increased yields as well as reduced need for water.

The committee found the following regarding the human health effects of GE foods:

> Many people are concerned that GE food consumption may lead to higher incidence of specific health problems including cancer, obesity, gastrointestinal tract illnesses, kidney disease, and disorders such as autism spectrum and allergies. In the absence of long-term, case-controlled studies to examine some hypotheses, the committee examined epidemiological datasets over time from the United States and Canada, where GE food has been consumed since the late 1990s, and similar datasets from the United Kingdom and western Europe, where GE food is not widely consumed. No pattern of differences was found among countries in specific health problems after the introduction of GE foods in the 1990s.[7]

There is no question that the GE technology should be closely regulated and the communication among researchers and the public enhanced. However, there is also no question that this technology has already demonstrated some significant benefits for both the environment as well as food production and availability, especially to the poor. As an example, vitamin A deficiency is the leading cause of preventable blindness in an estimated 500,000 children in underdeveloped countries such as the Philippines, India, and China, as well as contributing to the prevention of many other adverse health consequences of malnutrition. Now, what if you had a variety of rice that could be grown cheaply and had been genetically modified to produce more beta-carotene, a precursor to vitamin A, to help alleviate this problem. As a parent, would you resist its development or implementation, especially if your children were the ones at greatest risk?

The relatively new genetically engineered Golden Rice can now help solve this serious nutritional problem, as well as contribute to poverty reduction and food security in developing countries. However, as pointed out in her peer-reviewed paper "Benefits of Genetically Modified Crops for the Poor: Household

Income, Nutrition, and Health,"[8] Matin Qaim, Ph.D., from the Department of Agriculture Economics and Rural Development at Georg-August University in Germany, states the following:

> More public support is needed in biotechnology development, to ensure that other promising technologies for the poor are being developed, and in technology delivery, to ensure that they are widely accessible. In this respect, the negative public attitudes towards GM crops, especially in Europe, which are largely the result of biased information, are a fundamental obstacle. Not only do they limit public investments into GM crop research, but they also contribute to an overly complex regulatory framework. Some regulation is necessary to avoid risks, but over-regulation unnecessarily increases the cost of technologies, thus introducing a bias against small crops, small countries and small research organizations, which also implies a bias against the poor. This situation needs to be rectified through better and more science-based information flows.[9]

On April 13, 2010, the National Research Council of the National Academy of Sciences released its report, "Genetically engineered crops benefit many farmers, but the technology needs proper management to remain effective."[10] It points out that:

- Genetically engineered crops now constitute more than 80% of soybeans, corn, and cotton grown in the United States.
- Improvements in water quality could prove to be the largest single benefit of GE crops, due to the decreased need for insecticides and herbicides that may linger in the soil.
- Farmers who grow herbicide-resistant crops till less often to control weeds and are more likely to practice conservation tillage, which improves soil quality and water filtration and reduces erosion.

In 2011, Pamela Ronald, a UC Davis plant geneticist, pointed out some additional positive points in *Scientific American* in "Genetically Engineered Crops—What, How and Why":[11]

- By the turn of the century, the world population could increase to 10 billion.
- The amount of land and water is limited. We can no longer simply expand farmland to produce more food.
- We must develop new crop varieties tolerant of diverse stresses.
- There is broad scientific consensus that genetically engineered crops currently on the market are safe to eat. After 14 years of cultivation and a cumulative total of 2 billion acres planted, no adverse health or environmental effects have resulted.

- The economic benefits of reduced insecticide use costs carry over to non-GM crops, as well. This is due to the documented area-wide suppression of the primary pest which would affect non-GM crops as well in adjacent fields.
- GM crops improve beneficial insect biodiversity by allowing non-targeted beneficial insects to thrive.
- Corn and cotton have been genetically engineered to produce proteins from the soil bacterium Bacillus Thuringienis (Bt), which makes them resistant to some key caterpillar and beetle pests of these crops. Bt toxins cause little or no harm to most beneficial insects, wildlife, and people. Growers in Arizona who utilize this technology have been able to reduce insecticide use by 70%.

Ironically, the organic industry promotes its products as healthier for you. However, a report in the *Journal of Transgenic Research* points out:[12]

> Genetically modified (GM) Bt corn, through the pest protection that it confers, has lower levels of mycotoxins: toxic and carcinogenic chemicals produced as secondary metabolites of fungi that colonize crops. In some cases, the reduction of mycotoxins afforded by Bt corn is significant enough to have an economic impact, both in terms of domestic markets and international trade. In less developed countries where certain mycotoxins are significant contaminants of food, Bt corn adoption, by virtue of its mycotoxin reduction, may even improve human and animal health.

This is a result of the Bt developed corn's ability to destroy the corn borer insect before it has an opportunity to produce holes in the corn, which is the passageway for mold and the resulting development of mycotoxins.

If the general public, through ignorance and unfounded trust in deceptive anti-GE advertising, illogically rejects this technology and avoids those products which are produced as a result of GE, then seed and crop producers will have no financial incentive to develop it. It is very naïve to think that we should ignore or handcuff the development of technologies which will expand the production of safe food. There are well over 400 peer-reviewed papers from the scientific literature that document the safety and nutritional quality of GE foods. As illustrated in Chapter 9, this safety and accountability track record is far superior than any over-the-counter supplement that over 100 million American consumers blindly ingest each day. Most Americans blindly embrace supplements, but as Pew Research shows, a similar number of Americans reject a technology with extensive research demonstrating its benefits. This is all due to the profoundly one-sided negative narrative against GM foods.

The key for positive adaptation is to expose accurate information of the costs vs. benefits of a technology. A great example of this occurred in the fall 2013 semester after I had presented on the benefits of genetic modified foods to my basic nutrition class. We had met on a Monday night, and by the following Sunday

(September 15, 2013), I received the following email from one of my students: "This is an awesome article that made me think of our nutrition class topics." The student was referring to an article which had appeared in *Scientific American* on September 6, 2013, titled "Labels for GMO Foods Are a Bad Idea."[13] The article paralleled the information she received in class and reinforced her now better understanding of the issue through education vs. indoctrination. The article makes some points worth repeating here:

- If the California bill had passed, the average California family's yearly food bill would have risen by $400 with no benefit attached to it.
- Antagonism toward GMO foods also strengthen the stigma against a technology that has delivered enormous benefits to people in developing countries and promises far more. Recently published data from a seven-year study of Indian farmers show that those growing a genetically modified crop increased their yield per acre by 24 percent and boosted profits by 50 percent. These farmers were to buy more food—and food of greater nutritional value—for their families.
- At press time, GMO-label legislation is pending in at least 20 states. Such debates are about so much more than slapping ostensibly simple labels on our food to satisfy a segment of American consumers. Ultimately, we are deciding whether we will continue to develop an immensely beneficial technology or shun it based on unfounded fears.

Let me provide a very practical application of GE using an example provided by a Purdue University researcher[14] with tomatoes. About 20% of tomatoes are lost to spoilage due to short shelf life. Through genetic modification, a gene was able to be introduced which was able to modulate the ripening process and extend the tomatoes' post-harvest shelf life for about a week.

For consumers, this is obviously an advantage, especially for those who live in regions of the country where produce production may be limited and rely heavily on shipping to acquire what they need. This extended shelf life allows for greater options in the variety of food they can consume as well as support better health.

I recall in high school going to work in the tomato fields in Mettler, California, to pick tomatoes with high school friends for summer employment. At the time (1971), the pay was 25 cents per three-gallon bucket picked and delivered to the trailer. However, we knew that we could hustle through the fields fast enough and pick enough throughout the day to make it worthwhile. These wages were certainly more than minimum wage at the time, which was $1.65 an hour. Keep in mind that at the time, the cost of gas was only 25 cents per gallon and a brand-new Volkswagen Beetle only $1,900. So, our money went just a tad further back then, and thus, the pay scale was not the issue. The main issue was following the instructions of the field foreman and being certain you picked the green tomatoes and not the ripe red ones or your bucket would have no value to the grower and was dumped on the ground. In other words, you just wasted your effort. The tomatoes had to be green enough to be shipped for distribution. Ripe tomatoes would simply spoil prior to reaching their destination. So, by extending the shelf life of ripe tomatoes by one week can have a significant impact on consumers receiving field ripened tomatoes vs. green ones.

Additionally, through extensive hybridization, or old-fashioned genetic modification, there are now varieties of tomatoes with thicker skins that will allow them to be picked riper, as well as shipped farther. These advances obviously expand not only the market for tomatoes and increase the income for producers, but also the quality and quantity of food for greater numbers of consumers (sustainable agriculture). I would think that any reasonable person would consider this a good thing, not a bad thing. This is exactly the type of benefit GE technology can provide consumers.

Norman Borlaug, Ph.D., received the Nobel Peace Prize in 1970 for his lifetime work helping to feed the world's hungry. He is a Senior Scientist at the Rockefeller Foundation and a Distinguished Professor of International Agriculture, Department of Soil and Crop Sciences, at Texas A&M University. In an interview he had on Biotechnology and the Green Revolution, posted by AgBio-World, Dr. Borlaug made the following points:[15]

- Starting in the 1940s with funding from the Rockefeller Foundation, he was part of a program which took nearly 20 years to breed high-yield dwarf wheat that resisted a variety of plant pests and diseases and yielded two to three times more grain than traditional varieties so that poor farmers in Mexico could increase wheat production.
- They were then able to expand the program to poor farmers in Pakistan and India, which yielded an increase in wheat production from just 4.5 million tons in 1965 in Pakistan to 8.4 million tons just five years later in 1970. As for India, the production went from 12.3 million tons in 1965 to 20 million by 1970.
- In areas such as Africa where the cost of fertilizer is prohibitive due to the lack of transportation to get the fertilizers to farmers—no roads, no railroads, the advantages of GE crops can be even greater among rural farmers who combat a variety of crop-threatening issues due to the lack of technology to control crop losses, as well as production. You can bring GM seed to them more easily than you can fertilizer.
- Biotechnology will help these countries accomplish things that they could never do with conventional plant breeding. The technology is more precise, and farming becomes less time consuming.
- The technology allows us to have less impact on soil erosion, biodiversity, and wildlife due to loss of habitat which would have to be destroyed by conventional farming, forests, and grasslands to feed a world population which may reach 8.3 billion by 2025.

In 2012, California contained a ballot measure, Proposition 37, requiring the labeling of a variety of foods if they contained any ingredients from genetically modified crops, as briefly mentioned earlier in the *Scientific American* article. If it had passed, it would have cost consumers and manufactures billions of dollars with zero benefit. Fortunately, to my surprise, it failed. I was surprised because I was very aware of the level of consumer ignorance in this area and how easy it is to exploit this ignorance with the pseudoscience of those who wrote the initiative.

I must give credit here to a group which I am typically very cynical of, as illustrated in Chapter 2, but in this instance played a significant role in this defeat: the media. This is a case in which most of the print media in California got this issue right and did a good job of informing consumers of the pseudoscience and blatant fallacies used by the initiative's sponsors, as well as the cost California consumers would bear if it passed. More than 30 California newspapers' editorial departments were able to see through the initiative's nonsense and inform consumers to defeat the bill. However, I believe an equal role was played by those who opposed this bill and the effective public education campaign they implemented to defeat it. But the sad issue is, it literally took a $45 million barrage of educational advertising to help defeat it. Still, the margin of defeat, even after the newspaper editorials against it as well as the millions spent to defeat it, was only 53–47%. This is unfortunate, because it only reflects the extreme measures it takes to inform a science illiterate consumer that they are being fleeced and exploited by some ideological group of purported "consumer advocates."

Finally, a summary of this issue is provided by the University of Cornell's *Alliance For Science*, in "10 Myths About GMOs":[16]

> GMO stands for genetically modified organism. It most commonly refers to organisms—often plants—that have been modified to achieve desired traits, like drought-tolerance and pest-resistance, using recombinant DNA techniques or genetic engineering (GE). It's a misleading term, since we've been modifying the genetics of organisms since the dawn of agriculture. But the name isn't the only thing that people get wrong.

The report then goes on to provide the following 10 myths with their respective clarifications.

1. **Myth:** Farmers can't save GMO seeds.
   Reality: It is true that patented GMO seeds are often protected by intellectual property rules, meaning farmers must pledge not to save them and replant. Monsanto says it has sued about 150 farmers who it claims broke these rules over the past 20 years. However, hybrid seeds, which have been around for decades, also

need to be purchased each season because they don't breed true, so this is not a new issue for many farmers. In both cases, farmers choose to purchase these seeds because they get a better yield and make more money. In addition, in many public sector projects, such as the Hawaiian papaya, insect-resistant eggplant in Bangladesh, and the Water Efficient Maize for Africa partnership, farmers are free to save and share GMO seeds, and no royalties are charged.

2. **Myth:** GMOs are a corporate plot to control developing nations and the world's food supply.
   Reality: Developing nations are increasingly choosing GMOs, and for the fourth year in a row, devoted more hectares to growing biotech crops than developed nations. Farmers in these countries choose biotech because these crops have helped to alleviate hunger by increasing incomes for 18 million small-holder farm families, bringing financial stability to more than 65 million people in developing nations. This technology should not bypass the poor, who are arguably those who stand to benefit most.

3. **Myth:** GMOs are a ploy by agrichemical corporations to sell more pesticides/herbicides.
   Reality: Some GMO crops—such as Roundup Ready—can tolerate applications of herbicides, a trait that reduces the need for hand weeding or mechanical cultivation, which disturbs the soil. People sometimes imagine that GMOs use more insecticides, but the reverse is true with GMO crops that are bred with a natural form of insect resistance, thus minimizing or eliminating the need to spray pesticides for crop protection. Overall, scientists say GMOs have reduced the use of chemical pesticides—both herbicides and insecticides—by 37%.

4. **Myth:** GMOs are used only in industrial, chemical-intensive agriculture.
   Reality: The technology of genetic engineering can be used in multiple ways, including reducing pesticides. Today, many GMO crops are being bred in developing countries by public sector scientists who are working to improve the nutritional content and viability of staple food crops key to their region, such as cassava, pulses, mustard, brinjal, potatoes, rice, and bananas. Small-holder farmers typically grow these crops to feed their families.

5. **Myth:** GMOs are not adequately tested.
   Reality: Governments everywhere employ strict biosafety protocols to ensure that any new GM product poses no threat to human or animal health, or the environment. These protocols include

laboratory and field tests that may span many years. The resulting plants and foods are far more thoroughly tested than their conventional counterparts. Hundreds of scientific papers have assessed the safety of GM crops, and the vast majority found they are nutritionally equivalent to their conventional counterparts.

6. **Myth:** GMOs are harmful to the environment.

    Reality: Farmers who grow GMO commodity crops, like soy and corn, do less tilling, which reduces topsoil loss, erosion, and the associated runoff of fertilizer. They also can cultivate pest-resistant GMO crops, like Bt cotton, corn, and eggplant, with far fewer applications of pesticides, which benefits human and environmental health. Agriculture and its associated land use accounts for over a quarter of all global greenhouse gas emissions. On average, GE crops have reduced chemical pesticide use by 37%, increased crop yields by 22%, and increased farmer profits by 68%. GE crops also have reduced CO_2 emissions (mostly through enabling no-till farming practices) by 27 billion kg—equivalent to taking 12 million cars off the road.

7. **Myth:** GMOs are unhealthy.

    Reality: GMO foods have a long, safe track record during their more than 20 years on the market. The prestigious National Academies of Science agrees with US regulatory agencies, scientists, and leading health associations worldwide that food grown from GM crops is safe to eat, and no riskier than consuming the same foods containing ingredients from crop plants modified by conventional plant breeding techniques. Banning GMOs results in negative health consequences, because farmers would be forced to go back to using older, more toxic pesticides and access to food is more limited.

8. **Myth:** GMOs are unnatural.

    Reality: Humans have been selectively breeding plants and animals for countless millennia, so all domesticated plant species—and even your pet dogs and cats—are technically genetically modified.

9. **Myth:** Organics are safer than GMOs.

    Reality: Organic farming is a cultivation method, and GMOs are a breeding method, so it's like comparing apples and oranges. Additionally, organic growers are allowed to use certain types of pesticides, so some GMOs could claim to be safer than organics. An example might be a GM blight-resistant potato, which does not need toxic substances like copper sulfate or other fungicides often used to control blight in organic farming. Ideally, genetic modification would be used to improve organic farming.

10. **Myth:** GMOs won't feed the world.

Realty: No one plant breeding or agricultural system can or will feed 9 billion people in a sustainable manner. There is no "silver bullet." We need everything to help contribute to this goal: conventional, organic, biotech, small-holder, large-scale, as well as better distribution and storage systems, and less food waste, too.

In conclusion, I think most reasonable people will consider biotechnology or GE benefits as not only being good for the environment but humanity as well.

Let me highlight just a few of the positive points genetic engineering (GE) has to offer the food production process:

- Developing varieties which are more pest and pathogen resistant which will decrease the use of pesticides.
- Developing varieties which tolerate heat and drought.
- Reducing the acreage used for food production because yield per acre will be greater.
- Reducing soil erosion due to less tilling.
- Improving water quality due to a reduced need for insecticides and herbicides.
- Allowing the poor to eat by expanding food production in areas that were once less tolerant to a crop due to marginal soil or environmental factors.
- Increasing yield per acre at a decreased cost.
- Yielding nutritionally enhanced crops for regional use.
- Increasing shelf life of food.

Notes

1. https://www.pewresearch.org/science/2018/11/19/public-perspectives-on-food-risks/
2. Ibid, p. 2.
3. Alessandro Nicolia, Alberto Manzo, Fabio Veronesi, and Daniele Rosellini, An overview of the last 10 years of genetically engineered crop safety research. *Critical Reviews in Biotechnology* (2013),Vol. 34, No. 1, pp. 77–88.
4. www.hoover.org/research/rush-condemn-genetically-modified-crops
5. Board on Agriculture and Natural Resources, Division on Earth and Life Studies, *Genetically Engineered Crops: Experiences and Prospects – Report in Brief* (May 2006), p. 1.
6. www.nap.edu/resource/23395/GE-crops-report-brief.pdf
7. Ibid, p. 3.
8. M. Qaim, Benefits of genetically modified crops for the poor: household income, nutrition, and health. *New Biotechnology* (November 2010),Vol. 27, No. 5. pdfs.semanticscholar.org/3371/9e0fa633073472df2452d4f6df075e4f6516.pdf
9. Ibid, p. 5.
10. National Academy of Sciences. Genetically engineered crops benefit many farmers, but the technology needs proper management to remain effective, report suggests. *ScienceDaily* (April 22, 2010). www.sciencedaily.com/releases/2010/04/100413112058.htm
11. P. Ronald, Genetically engineered crops—what, how and why. *Scientific American* (2011).

12. F. Wu, Mycotoxin reduction in Bt corn: Potential economic, health, and regulatory impacts. *Transgenic Research* (June 15, 2006), Vol. 15, No. 3, pp. 277–89.

13. The editors at Scientific American, Labels for GMO foods are a bad idea, *Scientific American* (September 1, 2013).

14. www.purdue.edu/newsroom/research/2010/100628HandaTomato.html

15. www.agbioworld.org/biotech-info/topics/borlaug/bioscience.html

16. allianceforscience.cornell.edu/wp-content/uploads/2018/03/mythsFINAL.pdf

12
MISCELLANEOUS MYTHS AND MISINFORMATION

There are a plethora of myths, falsehoods, and erroneous assertions, which are meant to persuade consumers to adapt a certain lifestyle or purchase a product they neither have the need for or benefit from apart from possibly the placebo effect, and an emotional appeal. This chapter will cover in brief, a variety of assertions which are false but are commonly embraced as true by many consumers. Be wary of any manufacturer, product, or health guru that makes one of the following assertions.

The Need for Omega-3 Fatty Acid and Vitamin D Supplements

In 2018, the *New England Journal of Medicine* (NEJM) published the long-awaited Vitamin D and Omega-3 Trial (VITAL),[1],[2] which involved 25,871 healthy participants 50 and older who were followed for more than five years. The study was the largest study of its kind and the results would help define this area purported need of supplementation. The study was randomized so some participants received a placebo, while others received both the vitamin D and fish oil, and some either the fish oil or vitamin D alone. The main findings were as follows. Both trials were negative and not a surprise. Neither fish oil or vitamin D supplements lowered the incidence of heart disease and cancer. Why?

It is straightforward. For vitamin D, as I explained in Chapter 7, it would be difficult for most healthy adults who get any reasonable amount of sun exposure in addition to a reasonable diet, as well as the extensive reserves of vitamin D in fat tissue, to have such a low intake of vitamin D that a supplement would benefit them, which of course this study illustrates. Vitamin D supplements should not be consumed unless you have been diagnosed with a need to do so by a physician.

Now for the fish oil supplement. Let's examine the evidence for fish oil supplements and see if we cannot resolve this issue without the decimation of the fish population to supply misguided consumers with worthless fish oil supplements.

In addition to the most recent 2018 NEJM study previously mentioned, in May 2018, the *Journal of the American Medical Association Cardiology* (JAMA) published "Another Nail in the Coffin for Fish Oil Supplements."[3] The researchers reviewed 10 randomized trials, involving over 77,000 individuals, on the benefits of omega-3 fatty acid supplementation on the prevention of fatal or nonfatal cardiovascular events. The results: those individuals who consumed omega-3 fatty acid supplements failed to show any significant benefit of the supplements. This is consistent with the National Institute of Health summary of this issue in "Omega-3 Supplements: In Depth."[4] This report stated "Research indicates that omega-3 supplements don't reduce the risk of heart disease."

In 2016, the PBS *Frontline* investigative report "When It Comes to Supplements, What's Really in the Bottle?" questioned Adam Ismail, the spokesperson for GOED, a fish oil supplement trade association. *Frontline* simply asked him to produce the studies the fish oil supplement industry felt supported the benefits of fish oil supplements in the prevention of cardiovascular disease, stroke, or heart attack. He was unable to provide any evidence. In fact, the evidence he did provide, when reviewed by *Frontline*, illustrated that fish oil supplements were of no benefit (see deceptive marketing methods section in Chapter 1). *Frontline* also illustrated, with lab analysis of fish oil supplements, that many fish oil products contained oxidized lipids, which trigger inflammatory responses rather than prevent them.

First, it is certainly true that both omega-3 as well as omega-6 fatty acids are essential to your health for a wide variety of metabolic functions, and those whose diets are higher in omega-3 fatty acids from fish, nuts, vegetables, etc. (a good diet) appear less likely to experience strokes, heart attacks, cardiovascular deaths, etc. Common sense so far.

Second, it is also true that the lower ratio of 3:1 to 1:1 of these essential oils, compared to the 10:1 ratio or higher in most American diets, appear to be beneficial. But why? Is it due to some preconceived historical diet humans ate, or, is it simply related to the obvious—poor lifestyle choices, i.e. an association and not the cause-and-effect relationship marketers of fish oil supplements would like you to believe?

Answer: it is lifestyle related. Western diets are high in fast foods, French fries, chips, baked items, cookies, doughnuts, pastries, desserts of all kinds, etc., which contain vegetable oils loaded with—guess what? Yes, omega-6 fatty acids. On the other hand, this same population group consumes very little omega-3 rich foods such as nuts, fish, vegetables, etc. So, you have a population group with very poor dietary habits, which will naturally reflect the undesirable ratio of omega-6 to 3 fatty acid ratios, but this ratio simply illustrates abysmal overall dietary habits, which is why the omega-3 fatty acids supplements are of no benefit, as demonstrated by the research. So, if you wish to "optimize" your omega-6 to omega-3 ratios, and avoid the numerous diseases associated with it, simply make better food choices and improve your overall lifestyle. It's not the need for more omega-3s, but the decrease in the intake of the omega-6-rich foods—the baked and fried items.

Gluten-Free Diets

There is no question that those who suffer from celiac disease, as well as non-celiac gluten sensitivity (NCGS), benefit from a gluten-free diet, but just how much of the remaining marketplace is fabricated based upon deceptive marketing and the placebo effect? And, is gluten necessarily the offending ingredient?

In 2017, the journal *Clinical Gastroenterology and Hepatology* published "Suspected Nonceliac Gluten Sensitivity Confirmed in Few Patients After Gluten Challenge in Double-Blind, Placebo-Controlled Trials."[5] The authors reviewed ten previous studies in which individuals ate either a gluten-containing capsule or a placebo for a specified period and then switched, and neither the physician or the individual new the order. After each trail, gastrointestinal symptoms were evaluated. There were 231 NCGS individuals involved. The authors stated in the discussion section of the paper, "more than 80% of nonceliac patients, labeled as suffering from NCGS after a favorable response to a gluten-free diet, cannot reach a formal diagnosis of NCGS after a double-blind, placebo-controlled gluten challenge."[6] In the conclusion section of the report, the authors stated "the nocebo effect was detected in up to 40% of patients."[7]

So, what does this mean to those who embrace the gluten-free products?

1. Your response to gluten has nothing to do with the gluten, and your symptoms are related to another ingredient.
2. There is a bit of the nocebo effect taking place, which is the opposite of the placebo effect. The placebo effect precipitates a symptom from an inert substance which cannot cause the effect, while the nocebo effect causes you to feel better when you stop ingesting something, even though the ingredient would not have caused the symptoms in the first place.

The bottom line: don't go gluten free unless you have been properly diagnosed as having to do so. Going gluten free will not improve your health and will unnecessarily eliminate a variety of healthy foods from your diet. It is a fad for the majority of people who go gluten free.

Our Soils Are Depleted, So You Must Take Supplements

Agribusiness in any developed country has the enormous task of feeding billions of people. Doing this requires an exceptional amount of care, science, and technology to make sure that the soil is properly managed. Part of this management process includes testing and fertilization when necessary to be sure the soil has adequate minerals not only to grow a crop, but to obtain adequate yields per acre and quality of produce. Farming operations routinely conduct a soils test to evaluate soil conditions and to amend whatever is needed, as well as plant tissue petiole test. The petiole test examines the plants tissue to make sure the plant is taking up the appropriate nutrients from the soil. Plants require minerals for a variety of essential biological and physiological functions, so if adequate minerals were unavailable, as this myth purports, there would be a very noticeable—and likely unacceptable to most consumers—change in the physical structure of both the plant and the produce, as I illustrated with the few examples in Chapter 1.

Alkaline Foods or Water to Change Your pH

During the fourth week of a beginning nutrition course, body pH and how the body maintains a tight control of it, is covered in the course. Students learn why the body maintains a very limited range of pH and the various mechanisms it has at its disposal to do so. There is literally nothing you can consume by mouth which is going to change this. Alkaline water and foods and their relation to your body's pH is zero. This is basic chemistry. Your stomach's pH is 6.3 (low) for a reason. The enzymes which are present in your stomach to begin the digestive process of protein molecules work in that pH range. If the pH were to increase, this normal function of your stomach would be limited, as would your digestion of protein.

It Is Impossible to Eat Right, So You Need a Supplement

Even if this were true, chronic poor food choices—even with supplements—would leave you deficient in the thousands of plant compounds (phytochemicals) essential to good health. Due to advances in agricultural technology, food packaging, freezing, transportation and shipping, storage, etc. there has never been a time in history when the average consumer has had the luxury we now have in choosing from such a wide variety of quality nutrient-dense foods which are available year around. This issue is not "it is impossible to eat right"; the issue is poor consumer decisions and lifestyle habits. Poor personal accountability for your own health will not be resolved by any supplement. Just how many varieties of fresh fruits and vegetables, grains, etc., would someone have to have available to them before they think they have enough options?

Sugar Is Bad for You

In excess, yes, but the paranoia most have with sugar is as table sugar, and all the purported "studies" linking it to various diseases. Sugar is a 1-to-1 mixture of two smaller molecules of glucose and fructose, the same molecules found in all fruits, vegetables, and grains, which are readily metabolized by the body. So, the problem should have nothing to do with the molecule itself, but the quantity of it in the diet. As with any chemical compound, the *Principle of Toxicology—the dose makes the poison*, applies here, not the compound, as I have explained so many times in this book. It is common sense that excessive quantities of sugar, especially in inactive individuals, can result in excessive weight gain and the associated diseases which comes with obesity, but don't let the molecule itself instill fear in you. By itself, it is devoid of nutrient value, but sugar can certainly be utilized as an energy source. What can be more soothing than homemade chocolate chip cookies with milk?

Honey Is Better Than Sugar

It's not better, just different, depending on how you intend to use it. As an example, honey on a biscuit would be preferential than just table sugar due to the texture and taste differences, not the health differences. Ironically, many who embrace a paleolithic-related diet will embrace honey, which is made up mostly of glucose and fructose, the same two molecules found in table sugar, as well as HFCS. Honey has a minor number of various vitamins, minerals, other sugars and water, as well as other elements, but they are insignificant, and the end metabolic product of most of the honey, as table sugar, and HFCS, is glucose. The safety of HFCS is presented in Chapter 3. The point is, whether it is table sugar, HFCS, or honey, the end product of their metabolism, are the same.

Sugar Causes Hyperactivity

Variation in blood sugar levels do not cause hyperactivity. If this were true, we would all be in the mood for a good marathon run after any holiday feast where gluttony is the norm. What is taking place is simply a significant change in the mood of the child after the intake of the sugary tasting item caused by the hormonal changes from the anticipation of or receiving something sweet and pleasurable. Sugar is the association, not the cause. The hormonal changes are the cause. When these same children are provided a sweet-tasting drink made with an artificial sweetener, which would have no impact on blood sugar levels, their post "sugar high" and increased activities levels are the same. Any adult can directly relate to this immediate change in behavior when you are suddenly exposed to or informed of something which has made you happy. As an example, while watching any sporting event your immediate behavior can change dramatically from one moment to the next depending upon your mood, which is controlled by the situation or environmental factors, and not your blood sugar levels. In children, when the environmental factors change, as when they are handed a piece of chocolate cake, their attitudes—as well as behavior—will simultaneously change, as well.

Superfoods

This is one of the most annoying of all claims. They simply do not exist outside of cartoons such as Popeye and his spinach. This topic was already discussed in Chapter 1 under the Dr. Oz book review.

It Is Impossible to Obtain What You Need From Food Alone, So You Must Take a Supplement

I have already discussed the fallacy of attempting to apply the RDAs literally to each consumer daily. That is not the intention of these guidelines, as discussed. However, the supplement industry will attempt to misapply studies, which indicate that a large part of the population does not consume their daily intakes of these recommendations. It's irrelevant for most. The three mechanisms discussed—the storage capacity of all nutrients, the increased absorption rates in times of increased need, and the changes in retention rates in times of increase need—allow us to have a wide varied day-to-day intake and still maintain nutrient balance.

Modern Food Processing Destroys Most Nutrients

It is true that some of the methods used to process our food can destroy some nutrients. However, modern food processing increases the availability of nutrients in general to most by preventing spoilage, mold growth, and rancidity, and extending storage time. Millions live in locations where fresh fruit and vegetables would be very difficult to obtain on a year-round basis. Modern food processing provides most consumers with a wide variety of high-quality foods 12 months of the year—and the payoff is an increase in availability of nutrients overall. Modern food processing has allowed many who are reading this book to choose other employment endeavors other than growing your own food. Be thankful for modern food processing. Just choose wisely.

Modern Stress Increases Your Nutrients Needs

Stress is not a new phenomenon, and stress alone does not increase your nutrient needs. Much of our stress today is self-imposed and based on our wants, not our needs. In fact, the limited physical activity required for most lifestyles today has reduced our total caloric, and nutrient, need.

This Product Boosts Your Immune System

This issue is discussed in detail in Chapter 8. In short, most consumers seem to believe that their immune system is something they want to have "boosted" or "stimulated," when in fact, there is no physiological need for this. The body's immune system combats viral, fungal, and bacterial infections. If these infections are not present and your immune system is stimulated, what will it work to combat? Some researchers believe that the overstimulation of the immune system can lead to autoimmune diseases such as rheumatoid arthritis, diabetes, lupus, and multiple sclerosis. Keep in mind that through the body's natural adaptation to increased physical activity, athletes and active individuals already have an enhanced immune system.

Consume Only Locally Grown Food

I live in Kern County California, one of the most productive agricultural areas in the world, so I certainly have a wide variety of foods grown locally that I could survive on. However, this is not the case for most people. Most rely on the advances in preservatives and additives often demonized, as illustrated in Chapter 3, but which extend the shelf life and shipping distances of many products, allowing those living far removed from "locally" grown produce to eat in the first place. Locally grown certainly has its advantages, such as flavor or the sweetness of the produce, but it certainly is not the only option—nor would it be required for optimal health.

Notes

1. J. Manson, et al., Vitamin D supplements and prevention of cancer and cardiovascular disease. *New England Journal of Medicine* (January 3, 2019), Vol. 380, pp. 33–44.
2. J. Manson et al., Marine n-3 fatty acids and prevention of cardiovascular disease and cancer. *New England Journal of Medicine* (January 3, 2019), Vol. 380, pp. 23–32.
3. J. Abbasi, Another nail in the coffin for fish oil supplements. *JAMA* (May 8, 2018), Vol. 319, No. 18, pp. 1851–2.
4. S.M. Kwak, S. Myung, Y.J. Lee, H.G. Seo, Korean meta-analysis study group FT. efficacy of omega-3 fatty acid supplements (eicosapentaenoic acid and docosahexaenoic acid) in the secondary prevention of cardiovascular disease: a meta-analysis of randomized, double-blind, placebo-controlled trials. Archives of Internal Medicine (May 14, 2012), Vol. 172, No. 9, pp. 686–94.
5. Javier Molina-Infante, and Antonio Carroccio, Suspected nonceliac gluten sensitivity confirmed in few patients after gluten challenge in double-blind, placebo-controlled trials. *Clinical Gastroenterology and Hepatology* (2017), Vol. 15, pp. 339–48. www.cghjournal.org/article/S1542-3565(16)30547-X/pdf
6. Ibid, p. 344.
7. Ibid, p. 7.

INDEX

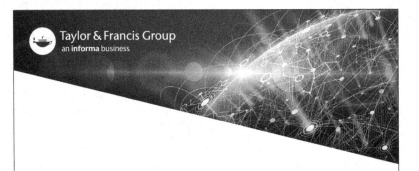

Printed in the United States
by Baker & Taylor Publisher Services